Solar Energy Today

Leslie F. Jesch

Solar Energy Laboratory
Birmingham University, UK.

Prepared for the
UK SECTION
of the
INTERNATIONAL SOLAR ENERGY SOCIETY

Leslie F. Jesch

SOLAR ENERGY TODAY

Published by:

UK-ISES

19 Albemarle Street, London.
W1X 3HA, UK.
Tel: 01 493 6601

The opinions expressed in this book are those of the author and do not necessarily reflect UK-ISES policy. Photographic and written references do not imply that the design, installation or product is approved by the Society, they serve only as illustrations of contemporary practice.

ISBN 0-904963-27-6

Printed by Bridgend Printing Co. Ltd., Tremains Road, Bridgend, Mid Glam.

Front cover photo: Arthur Burgess

PREFACE

In 1976 UK-ISES published 'Solar Energy, a UK assessment', which was a major review of solar energy and gave particular emphasis to the potential contribution which solar energy might make to the future energy requirements of the UK. Over the past few years a great deal of new information has emerged, partly as a result of the increase in national and international research and development programmes. There has also been a marked increase in the number of demonstration projects and in the general level of awareness of renewable technologies. It is therefore timely that UK-ISES should bring out a major new publication. 'Solar Energy Today' looks at the latest trends in solar technologies and also includes a chapter on wind energy which was not featured in the first publication. The final chapter illustrates by examples drawn from the UK the wide range of technologies currently being pursued.

B. Justin
Chairman. UK-ISES
August, 1981.

ACKNOWLEDGEMENT

When the Committee of the UK-ISES commissioned me to write a book on solar energy I realised that the available time made the task impossible without the help of my colleagues. I turned to the solar community and received generous help from specialists in many areas. I want to thank J. L. J. Rosenfeld, F. C. Treble, J. Coombs and B. Justin who helped with renewable energy sources, photovoltaic solar cells, biomass and applications. K. F. Jesch helped with the references, name index and subject index. W. B. Gillett helped with proof reading and printing matters. I had very useful discussions with Justin, Rosenfeld and Gillett which helped in deciding the policy of organisation and the style in which the book was written. Their time and suggestions were given generously and are much appreciated. I also received information, references, photographs and drawings from individuals and organisations. Their contribution is acknowledged under credits listed at the end of the text.

L. F. Jesch
Solar Energy Laboratory
August, 1981.

TABLE OF CONTENTS

Chapter 1

RENEWABLE ENERGY SOURCES

Compared to the known and estimated reserves of fossil fuels and uranium, the energy potentially available from renewable sources is enormous. The solar energy falling just on the land surface of the earth each year equals about four times the total estimated fossil fuel resources, and about four thousand times humanity's current global energy consumption. Wind, the other renewable energy source considered in this book, could also provide more energy than we now consume, despite the fact that only a small portion of it is available close to the earth's surface. Other renewable energy sources, whether derivates of solar energy such as hydropower, waves, ocean thermal gradients and currents, or the result of other forces such as tides, do not have quite the same potential on a global scale, but are or could be very significant in favoured regions.

Before the increase in the intensity of energy use which accompanies the industrial revolution, renewable sources supplied all mankind's energy needs. Even today about 25% of energy used comes from renewable sources such as hydropower, wood, dung, wind power and animal traction. The greater part of this is often omitted from energy statistics because it does not form part of the market economy. To increase the use made of renewable energy sources would therefore be to return to a tradition, rather than to break with one. The differences with the past will lie in the technology by which these sources are tapped and in the scale required to suit contemporary needs.

1.1 The Physical Nature of Solar Energy

The radiant flux intercepted by a surface facing the sun just outside the earth's atmosphere when the earth-sun distance is at its average value, is called the solar constant. The classical measurements carried out by the Smithsonian Institution in the first quarter of the century and described by Abbott (1922) have only recently been overtaken by the advent of high flying aircraft and satellites which have permitted measurements free from atmospheric interference. The recent review by White (1977) has settled on a value of 1.373 kWm^{-2} for the solar constant. The elliptical orbit of the earth causes a seasonal variation of about ±3%, the maximum occurring in early January and the minimum in July. There appears to be a slight dependence of the solar constant on the number of sunspots and solar flares cause small fluctuations of short duration, especially in the X-ray region of the spectrum. For all practical purposes however these effects may be ignored.

The spectrum of solar radiation outside the earth's atmosphere closely approximates to that of a black body at 5760 K. It is shown in Figure 1.1.1. Detailed measurements were made in the late 1960's from an aircraft flying at 12 km altitude and from a Mariner Mars probe. Thekaekara (1973) has published tables of the energy distribution as a function of wavelength. More than 99% of the spectrum lies between 0.23 and 4 μm. About 9% of the energy is in the ultraviolet, below 0.4 μm. Some 40% of the energy is in the visible part, between 0.4 and 0.72 μm and the remaining 50% lies in

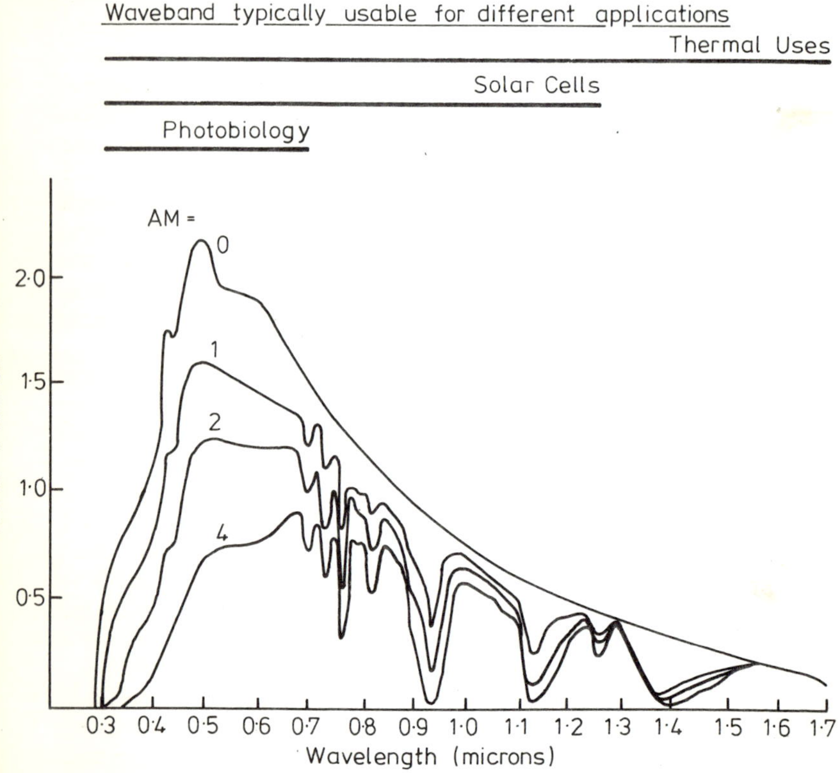

Figure 1.1.1: Solar radiation versus wavelength. Spectral distribution for different values of air mass: AM= 0,1,2,4.

the infrared, beyond 0.72 μm. These data are accepted as standard by ASTM (1974) and the IEA (1980).

As the solar radiation traverses the atmosphere several complex physical and chemical effects occur. Details may be found in text books and extensive reviews for example Sorenson (1979), Thekaekara (1977). Here only a brief summary is presented. Even in a clear sky, about 6% of the radiation is reflected by the atmospheric gases, dust particles and aerosols. Most of the ultraviolet light is absorbed in the upper atmosphere, principally by ozone and oxygen molecules. The infrared radiation is partly absorbed at all wavelengths by aerosols and dust particles and much more strongly absorbed over certain wavelength bands by carbon dioxide, water vapour and other polyatomic molecules. Radiation scattering by gas molecules (Rayleigh scattering) and by small aerosols and dust particles (Mie scattering) reduce the direct or beam component of the solar irradiance at ground level and give rise to the luminance of the sky, which is seen at the ground as the diffuse component of the solar irradiance. The amount of light absorbed and scattered depends on the composition of the atmosphere, the main variables being the turbidity (related to the concentration of aerosols and dust) and the water vapour concentration. In addition, the optical path length measured in terms of the air-mass is of course important. This is equal to one at sea level when the sun is directly overhead. If the earth was flat and the atmosphere uniform, the airmass would be equal to the cosecant of the sun's elevation. This is a very good approximation at elevations greater than ten degrees. For smaller elevations Kasten (1966) has presented a formula which allows for the curvature of the earth and refraction effects. A correction must also be made for variations in atmospheric pressure. In the UK, at latitudes between 51° and 55°N, the airmass at noon varies between about 1.1 in the summer and 4 in the winter.

From a knowledge of the absorption and scattering properties of the different components of the atmosphere, it is possible therefore to calculate the direct and diffuse irradiance at ground level, and their spectral distributions. Thekaekara (1977) and Leckner (1978) give results for different values of turbidity and air mass. A previously very often used set of spectral distributions is that published by Moon (1940). Figure 1.1.1 shows the spectral distributions for different air masses. At air mass 1, total irradiance on a clear day is reduced to about 960 Wm^{-2}, with about 5% in the ultraviolet, 37% in the visible and 48% in the infrared part of the spectrum. At airmass 4, the corresponding values are 600 Wm^{-2}, 1%, 44% and 55%.

The spatial distribution of the diffuse irradiance in a clear sky resulting from Rayleigh scattering is approximately isotropic, but the Mie scattering from aerosols produces a forward scattered circumsolar component, whose intensity decreases exponentially away from the sun. There is also an increase in irradiance from the sky close to the horizon. Although it is often assumed that the sky irradiance is isotropic, this can lead to serious errors, especially when the irradiance on inclined surfaces is to be estimated. Recently, Hooper (1980) and Brunger have described a model for clear sky irradiance which treats the three components explicitly to yield the ratio of the diffuse component on an inclined surface to that on a horizontal surface. The three coefficients which appear in their model were fitted to measurements of sky irradiance at Toronto, Canada. Their values depend on the sun's elevation and their dependence on turbidity was not tested. Steven (1979) and Unsworth and also Bodmann (1980) and Burger have shown that the forward scattered component can be treated as part of the direct irradiance and the other two components combined into an azimuthally independent function which increases towards the horizon. The two parameters in their models have been fitted to measurements of sky irradiance in the UK and Germany and have been found to be insensitive to turbidity and solar elevations over a wide range.

Clouds reflect, attenuate and scatter the solar radiation. In a partly clouded sky the effect on the irradiance at ground level cannot be exactly calculated as it depends on the nature and height of the clouds and their position relative to the sun. On average clouds reflect about 24% of the solar radiation back out into space. For a completely overcast sky the direct irradiance is of course zero, but the spatial distribution of the diffuse irradiance still shows some anisotrophy which can be expressed by a simple one parameter function as described by Steven (1979) and Unsworth. The model of Hooper (1980) and Brunger also fits the data for overcast skies.

1.2 **Measurements and estimation of solar irradiance**

Most instruments used in the measurement of solar irradiance are thermopiles in which a voltage proportional to the irradiance is generated between the 'hot' junctions which are either in contact with a heat sink shielded from the light or are painted white to reflect it. Convective heat losses are suppressed by covering the sensor with one or two transparent domes. Although photovoltaic cells are sometimes used they are not recommended for accurate measurements because their sensitivity depends on the wavelengths of the light and the spectral distribution of solar radiation at ground level varies with air mass and meteorological conditions.

To measure the direct component a narrow angle pyrheliometer can be mounted on a tracking mechanism to follow the sun. To measure the global (direct plus diffuse) irradiance, a 180° pyranometer is used. The diffuse irradiance can be measured separately by shading the pyranometer from the direct component, either with a disc which tracks the sun, or with a shading ring. The latter method is much simpler to use, but as discussed by Le Baron (1980) and co-workers and by Painter (1981) care must be taken to correct for the obscuration of part of the sky by the shade ring. The anisotropy of sky irradiance is particularly difficult to correct for.

The World Meteorological Organisation (1968) has defined the properties required for different classes of instruments according to the accuracy and precision needed for different applications. Primary standard instruments are kept at the World Radiation Centre at Davos in Switzerland. One important characteristic of a pyranometer is its ability to respond with equal sensitivity to radiation coming from any direction. The output from an instrument with uniform response will be proportional to the cosine of the angle between the incoming radiation and the normal to the sensor. Deviations from the cosine response are the main cause of uncertainty in the use of existing instruments. There are active programmes under the aegis of both the CED and IEA to measure the deviations from cosine response and to intercompare instruments used in different countries. According to Gillett (1980) and McGregor the instruments used in the European network have an error of about ±5%, rising to as much as ±15% for radiation coming from low altitudes. The status of the IEA programme has been reviewed by Richer (1981) and Stoffel. Mohr (1979) and co-workers describe the calibration procedure for some pyranometers deployed in the United States. A result common to all these studies is that

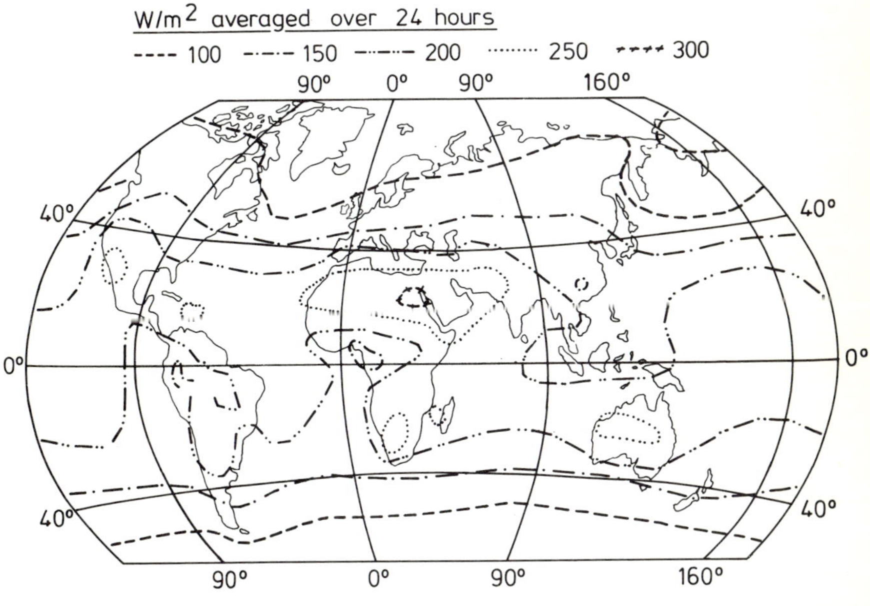

Figure 1.2.1: Mean annual distribution of daily solar radiation.

each instrument has its individual response characteristics and must therefore be separately calibrated.

The mean annual distribution of solar radiation reaching the earth's surface is shown in Figure 1.2.1. As can be seen, the annual isolation is mainly a function of latitude. In the higher inhabited latitudes of Northern Europe, the solar energy falling on one m^2 in a year is about 3.3 GJ (equivalent to 120 kg of coal). In the areas of maximum insolation, such as the Red Sea, the annual total is only about 2.5 times greater. If one considers only the regions between the 45° latitudes, within which 90% of the world's population lives, the variation is only twofold. The annual distribution of radiation does not adequately define its availability as an energy source, because it fails to take account of seasonal variations. This is particularly important at high latitude where the peak in energy demand occurs in winter when the solar radiation is weakest. Terjung (1970) devised a two dimensional classification of the global energy resource which takes the seasonal variation into account. When considering the use of plants as a means of collecting solar energy, multidimensional classifications are needed, which also take into account temperature and rainfall. The climate classification of Koppen (1936) is often used.

De Jong (1973) has published world maps of monthly mean global irradiance, based on the earlier work of Lof (1965) and co-workers. Although the sparsity of data in many parts of the world necessitated some guesswork and long interpolations, these maps are still very useful for preliminary evaluations. Increasingly countries are setting up networks of stations to measure solar irradiance. On behalf of the World Meteorological Service, the Russian Meteorological Office in Leningrad collects data from all over the world and publishes it in a monthly journal, Solar Radiation for the World. Unfortunately it is still quite difficult to extract a reliable long term record of radiation data for any particular region. More recently national or regional maps and statistical analyses have been published. Good examples are those of the UK Meteorological Office (1980) and of the CEC edited by Palz (1979). Data are also becoming available on magnetic tape or as computer data-banks, for example the SOLMET system in the United States.

For a detailed evaluation of the performance of solar systems it is often useful to know the direct and diffuse irradiance separately, preferably averaged over intervals short compared to a day. The commonest time interval at which data are recorded is one hour. However there are still many stations where only the global irradiance is measured, and often only the daily average recorded. Several equations, based on empirical correlations have been developed to overcome the deficiencies in the records. The ratio of diffuse to global irradiance has been correlated with the ratio of global to extraterrestrial radiation by several authors. A commonly used correlation for estimating diffuse (hence also the direct) irradiance from a record of global radiation is that of Liu (1960) and Jordan. An alternative correlation was proposed by Page (1961). Powell (1979) and Yellott reported that Page's equation gave better results. Other models have been proposed and Erbs (1980) has summarised and compared the various formulae. Their accuracy decreases and their variance increases as the time interval is reduced. Thus the estimate of mean monthly diffuse irradiance from the mean monthly global irradiance is quite accurate, but for hourly estimates large errors can occur. However since the long term average is good, the effect of such errors on the evaluation of system performance is not too serious.

Another parameter for which many records exist is the number of sunshine hours per day. Angstrom (1924) was the first to propose a correlation between the daily global radiation, the number of sunshine hours and the maximum possible values of these quantities. Page (1976) and more recently Iqbal (1979) have also considered such correlations. It appears that the coefficient in the correlations vary from site to site and De Jong (1973) lists them for many different parts of the world.

Another approach is to generate sequences of values averaged over short intervals (hourly or daily) from longer term averages using a stochastic model. Brinkworth (1977) has shown that provided a proper account is taken of the autocorrelation which exists in sequences of real irradiance records, it is possible to generate sequences of daily global irradiance which vary randomly about the long term mean in a realistic manner. Mustacchi (1979) and co-workers describe a stochastic model to generate sequences of hourly global irradiance.

Finally, Aydinti (1980) and Page (1979, 1981) have published the methods now adopted by the CEC to calculate the hourly average direct and diffuse irradiance on an arbitrarily inclined surface for clear and overcast days, and on intermediate days by weighting the two extremes according to the relative duration of sunshine. These methods are based on the atmospheric models described in section 1.1. The turbidity is defined by the Links turbidity factor. Although this is based on a Rayleigh scattering model of the atmosphere, and is to that extent less realistic than the Angstrom turbidity model often used in the American literature, it can be simply obtained from a comparison of the direct irradiance with the calculated extra terrestrial irradiance. One can therefore integrate the scattering over all wavelengths and also allow for the water content of the atmosphere. Because of its inherent derivation, the Links turbidity factor does vary with location, time of day and season, but these effects can be incorporated in the models. Values of the mean Links turbidity coefficients for various types of location are listed, for example by Rattunde (1980).

1.3 **Long Wave Radiation Exchange**

The earth emits black body radiation at infra red wavelengths greater than about 3 μm, i.e. at wavelengths longer than most of the solar radiation. This long wave radiation is partially absorbed by water vapour, carbon dioxide and clouds in the sky and re-emitted downwards with a spectral distribution corresponding to some mean sky temperature. There is thus an exchange of long wave radiation between the sky and the ground, or between the sky and, for example, a solar thermal collector or a building. In general the net balance causes a loss of heat from the ground to the sky, especially in clear, dry conditions. This is the cause of ground frost which can damage plants, and affects the performance of solar thermal systems.

The radiation exchange can be treated as that between two grey bodies. For this purpose one may either assume that the effective emittance of the sky is one and define its downward radiating properties in terms of an effective sky temperature, or one may assume that the sky is at the ground level ambient temperature and characterise its radiation by an effective emittance. Either the sky temperature or its effective emittance can be related to the water vapour content and clouds in the atmosphere. Bliss (1961) showed that in clear skies the important parameter is the water content of the first few hundred metres above the ground and he related the sky's emittance to the humidity measured at ground level. Such a simple formula does not apply to cloudy skies. De Jong (1973) summarises the correlations which have been attempted with cloud index, but account must be taken of the type and height of the clouds.

The long wave radiation from the sky can be measured using a pyrgeometer or a pyradiometer (which measures both infrared and visible radiation, the latter having to be measured separately and subtracted). These instruments are thermopiles, with either convection suppression as for pyranometers, or with forced convection for which allowance can be made. The design of modern instruments is discussed in the IEA (1980) handbook. Cooper (1981) and co-workers have recently published details of a new instrument which uses forced convection across a heat flow meter and illustrate its use in Australian conditions.

Measurements of the long wave sky radiation at Lake Ontario, Canada, have been used by Atwater (1978) and Ball to fit a model which takes into account absorption by water vapour and carbon dioxide as well as partial cloud cover at different heights. Typically long wave sky radiation varies between 100 and 500 Wm^{-2} depending on weather conditions. They also express their results in terms of effective sky temperature and show seasonally averaged maps for the United States. The effective sky temperature is usually between 6 and 18°C colder than ground level air temperature. Generally speaking therefore a solar thermal collector operating at or above ambient temperature will lose more long wave radiation than it receives. Green (1979) and Gillett have shown that this must be taken into account when testing the thermal performance of collectors, especially when comparing outside tests with those done in a solar simulator.

1.4 **Wind Energy**

The wind is an indirect manifestation of solar radiation and has in itself long been an important renewable energy resource. The total kinetic energy in the atmospheric wind system is huge, but most of this is in the zonal jetstreams which occur at mid latitudes about 10 km above the ground. Of more practical interest is the wind speed and the potentially available power close to the earth's surface. Figure 1.4.1 shows the world distribution of mean wind speed about 10 m above ground level for January and July. These maps do not take account of the effects of local topography in either creating sheltered areas or in causing a local acceleration which results in a windy area in otherwise calm surroundings. Nevertheless it is possible to draw some general conclusions. Areas of high wind occur predominantly over the oceans, and with few exceptions the only land based regions of high wind are along the coast. Even so, the potentially available power, which goes as the cube of the wind speed, within reach of

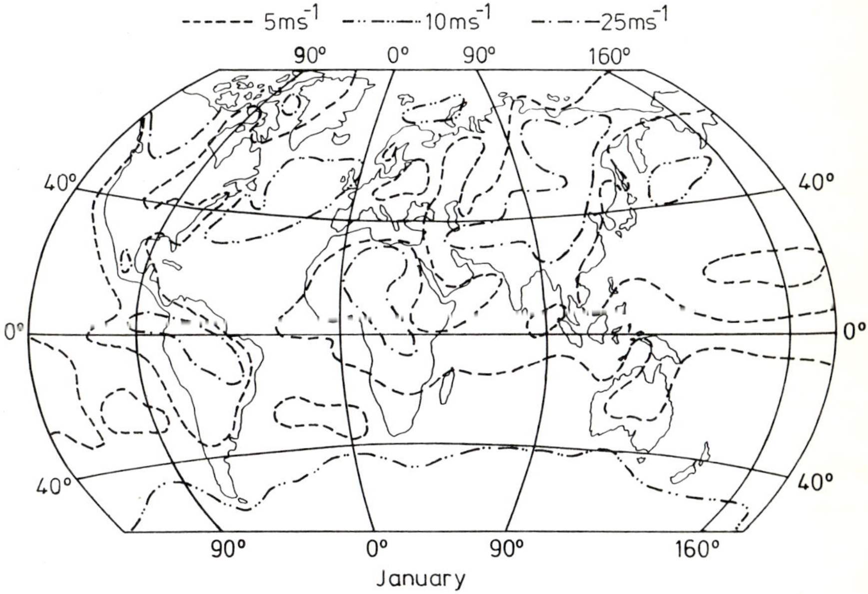

Figure 1.4.1.A: Annual average wind velocity distribution at 10m height in January.

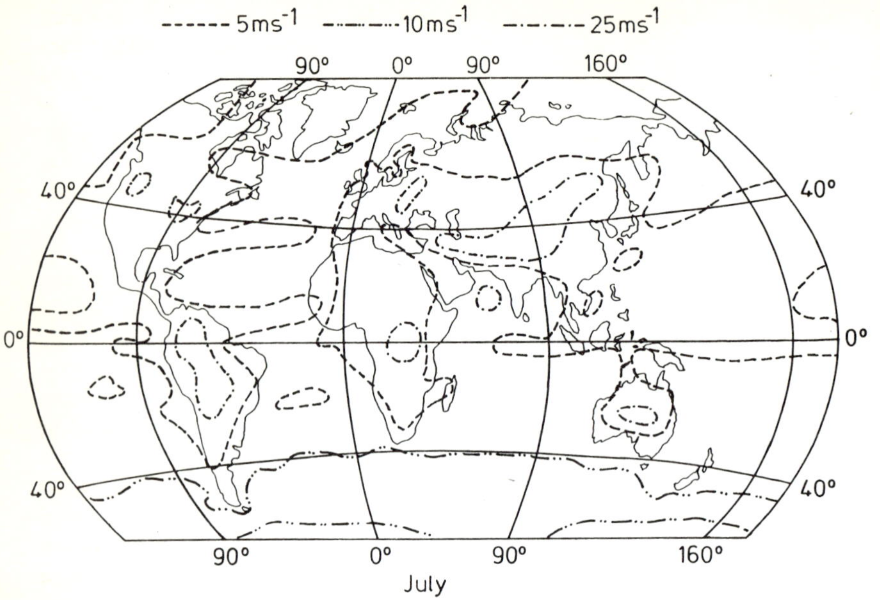

Figure 1.4.1.B: Annual average wind velocity distribution at 10m height in July.

man-made structure, probably lies in the region of 10^{12} to 10^{13} W, about the same as the current global rate of energy consumption. Although the seasonal variation in wind energy is less than for solar radiation, it is noteworthy that for high latitudes the winds are generally stronger in the winter, which in contrast to solar energy, matches the seasonal energy demand pattern. This suggests that for regions such as north western Europe and the Eastern parts of North America, solar and wind energy could be complementary sources.

Another important characteristic of wind is the strong variation of wind speed with height near the ground. The precise nature of the vertical wind speed profile depends on the local terrain and on weather conditions, but the general pattern is that the wind speed increases rapidly in the first 10 to 15 metres, then more gradually, but may not reach its asymptotic value until 100 m or more above the ground. It is therefore important, when assessing the suitability of a site for a wind machine, to measure the wind at the appropriate height. Modern large wind turbines have blades that sweep through the air between 50 and 150 m above the ground. Hence the blade will experience different wind speeds at the bottom and top of its rotation.

Because of the low density of air, the power density of wind is also very low. At a wind speed of 5 ms^{-1}, close to the mean wind speed in many inland areas, the available power is about 80 Wm^{-2} measured for an area perpendicular to the wind. Allowing for the fact that the maximum extraction efficiency of a turbine is about 60% (the Betz criterion), it would require about 22 m^2, or a circle of 5.2 m diameter to extract 1 kW. Of course, because of its dependence on the cube of the wind speed the available power increases rapidly in stronger winds, at 10 m^{-1}, for example, a circle of 1.9 m diameter is sufficient to extract 1 kW. As mentioned earlier, the local topography can drastically affect the wind. It is therefore always necessary, even in a generally windy area, to choose the site of a wind machine with great care.

Actual wind data are only available for a few scattered sites around the world, but mean wind speeds have been recorded in many places. To evaluate the performance of a wind machine it is however important to take account of the short term fluctuations in wind speed. A stochastic model has been developed by Chou (1981) and Corotis to generate a series of hourly wind speeds, given the monthly mean speed and its variance.

Page (1979) and co-workers have developed correlations based on a Fourier analysis of UK weather records, which relates the mean hourly wind speed with temperature and solar radiation for different classes of radiation days. (10% maximum, average and 10% minimum radiation design days). This is particularly useful to estimate the effect of wind on the thermal performance of buildings, but to apply the method elsewhere would of course require a full analysis of local weather data.

References

Abbott, C. G. (1922): "Results of the measurements of the intensity of solar radiation." **Annals of the Astrophysical Observatory of the Smithsonian Institution 4** 1922, 193-195.

Angstrom, A. K. (1924): "Solar and terrestrial radiation." **Quarterly Journal of the Royal Meteorological Society 50** 1924, 121-125.

ASTM (1974): "Standard specification for solar constant and airmass zero solar spectral irradiance." ASTM Standard E 490-73a. **Annual Book of ASTM Standards Part 41,** ASTM 1974.

Atwater, M. A. (1978) and J. T. Ball: "Computation of I.R. sky temperature and comparison with surface temperature." **Solar Energy 21** 1978, 211-216.

Aydinti, S. (1980): "The calculation of the available solar radiation and daylight." **Proceedings of the Symposium on Daylight,** Berlin. ed. J. Krochmann. Commission Internationale de L'Eclairage 1980, 23-32.

Bliss, R. W. (1961): "Atmospheric radiation near the surface of the ground: a summary for engineers." **Solar Energy 5** 1961, 103.

Bodmann, H. W. (1980) and W. Burger: "Irradiance on inclined surfaces with turbid skies." **Proceedings of the Symposium on Daylight,** Berlin. ed. J. Krochmann. Commission Internationale de L'Edlairage 1980, 33-41.

Brinkworth, B. J. (1977): "Autocorrelation and stochastic modelling of insolation sequences." **Solar Energy 19** 1977, 343-348.

Chou, K. C. (1981) and R. B. Corotis: "Simulation of hourly wind speed and array wind power." **Solar Energy 26** 1981, 199.

Cooper, P. I. (1981), E. A. Christie and R. V. Dunkle: "A method of measuring sky temperature." **Solar Energy 26** 1981, 153-158.

de Jong, B. (1973): **Net Radiation Received by a Horizontal Surface at the Earth.** Delft University Press 1973.

Erbs, D. G. (1980): **Methods of Estimating the Diffuse Fraction of Hourly, Daily and Monthly-Average Global Solar Radiation.** MSc Thesis, University of Wisconsin, Madison 1980.

Gillett, W. B. (1980) and J. McGregor: **Report of the CEC Pyranometer Workshop, Brussels, September 1980.** SEU Report 223. Solar Energy Unit, University College, Cardiff, UK. 1980.

Green A. A. (1979) and W. B. Gillett: "The significance of longwave radiation in flat plate solar collector testing." **Future Energy Concepts.** IEE Conference Publication No. 171, London 1979, 36-39.

Hooper, F. C. (1980) and A. P. Brunger: "A model for the angular distribution of sky irradiance." **Transactions of the ASME, Journal of Solar Energy Engineering 102,** 1980, 196-202.

IEA (1980): **An Introduction to Meteorological Measurements and Data Handling for Solar Energy Applications.** Document No. DOE/ER-0084, US Department of Energy October 1980. International Energy Agency Programme, Task IV.

Iqbal, M. (1979): "Correlation of average diffuse and beam radiation with hours of bright sunshine." **Solar Energy 23** 1979, 169.

Kasten, F. (1966): "A new table and approximation formula for the relative optical air mass." **Archiv fur Meteorologie, Geophysik und Bioklimatologie, Ser. B** 1966, 14.

Koppen, W. (1936): "Der geographische System der Klima." in **Handbuch der Klimatologie,** ed. W. Koppen and R. Geiger, Borntrager, Berlin 1936, Vol. 1.

Le Baron, B. A. (1980), W. A. Peterson and I. Dirmhirn: "Corrections for diffuse irradiance measured with Shadowbands." **Solar Energy 25** 1980, 1-13.

Leckner, B. (1978): "The spectral distribution of solar radiation at the earth's surface – elements of a model." **Solar Energy 20** 1978, 143.

Lof, G. O. G. (1965), J. Duffie and C. O. Smith: **World Distribution of Solar Radiation.** Engineering Station Report 21, University of Wisconsin, Madison 1965.

Liu, B. Y. N. (1960) and R. C. Jordan: "The inter-relationship and characteristic distributions of direct, diffuse and total solar radiation." **Solar Energy 4** 1960, 1-19.

Mohr, A. J. (1979), D. A. Dahlberg and I. Dirmhirn: "Experiences with tests and calibrations of pyranometers for a mesoscale solarirradiance network." **Solar Energy 22** 1979, 197-203.

Moon, P. (1940): "Proposed standard radiation curves for engineering use." **Journal of the Franklin Institute 230** (5) 1940, 583.

Mustacchi, C. (1979), V. Cena and M. Rocchi: "Stochastic simulations of hourly global radiation sequences." **Solar Energy 23** 1979, 47-51.

Page, J. K. (1961): "The estimation of monthly mean values of daily short wave radiation on vertical and inclined surfaces from sunshine records for latitudes 40°N−40°S. Working paper No. E/Conf. 35/5/81 **United Nations Conference on new sources of energy,** Rome, May 1961.

Page, J. K. (1976): **The Estimation of Monthly Mean Values of Daily Shortwave Irradiation on Vertical and Inclined Surfaces from Sunshine Records for Latitudes 60°N−40°S.** Building Science Interim Report BS 32. Dept. of Building Science, Sheffield University 1976.

Page, J. K. (1979): "Methods for the estimation of solar energy on vertical and inclined surfaces." in **Solar Energy Conversion, an Introductory Course.** ed. A. E. Dixon and J. D. Leslie, Pergamon Press 1979, 37-39.

Page, J. K. (1979), C. G. Souster and S. Sharples: "Mathematical modelling of hourly variations in temperature, wind speed and long wave radiation for different classes of radiation day in the U.K." **Meteorology for Solar Energy Applications** Conference (C18) London, UK-ISES 1979, 108-126.

Page, J. K. (1981): **Final Report on the Predeterminations of Irradiation on Inclined Surfaces for Different European Centres.** Solar Energy Programme, Project F. Solar Radiation Data, Action 3.2, Directorate General XII, CEC Brussels 1981, in press.

Painter, M. E. (1981): "The shade ring correction for diffuse irradiance measurements." **Solar Energy 26** 1981, 361.

Palz, W. (1979) ed.: **European Solar Atlas Vol. 1 Global Radiation on Horizontal Surfaces.** Commission of the European Communities, Brussels 1979.

Powell, G. L. (1979) and J. I. Yellot: "A procedure for determining solar heat gain factors for south-facing vertical surfaces on average days. **Proc. ISES Silver Jubilee Congress in Atlanta, Vol. 2** pp 1637-1641.

Rattunde, R. (1980): **Optimierung der Tageslichtbeleuchtung Grosser Raume durch Oberlichter unter Berucksichtigung des zur Verfugung stehenden Tageslichtes.** Doctoral Thesis, Technical University, Berlin 1980.

Richer, M. R. (1981) and T. L. Stoffel: **Pyranometer Measurements.** International Energy Agency Conference, Boulder, March 1981. Report No. SERI1TR-642-1156, Solar Energy Research Institute, Golden Colorado, 1981.

Sorenson, B. (1979): **Renewable Energy,** Academic Press 1979.

Steven, M. D. (1979) and M. D. Unsworth: "The radiance distribution of clear and overcast skies." **Meteorology for Solar Energy Applications.** Conference (C18) London, UK-ISES, 1979, 26-36.

Terjung, W. H. (1970): "A global classification of solar radiation." **Solar Energy, 13,** 1970, 67-81.

Thekaekara, M. P. (1973): "Solar energy outside the earth's atmosphere." **Solar Energy 14** 1973, 109-127.

Thekaekara, M. P. (1977): "Solar irradiance, total and spectral." **Solar Energy Engineering.** ed. A. A. B. Sayigh, Academic Press 1977, Chapter 3.

U.K. Meteorological Office (1980): **Solar Radiation Data for the U.K. 1951-75. Met. O.912.** UK Meteorological Office, Bracknell 1980.

White, O. R. (1977) ed.: **The Solar Output and its Variation.** Colorado Associated University Press 1977.

World Meteorological Organization (1968): **Guide to Meteorological Instruments and Observing Practices,** 3rd ed. W.M.O. Geneva 1968.

Chapter 2

PASSIVE SOLAR SYSTEMS

One of the earliest and certainly the most enduring applications of solar energy is found in the way we design, build and use the houses we live in. The whole thinking about houses and buildings has been so pervaded by solar energy utilisation concepts that most people would not even recognise that they are thinking solar when they talk about a suntrap verandah, a conservatory, or windows on some walls but not on others in a house. Estate agents use the expression 'southern aspect' to imply a desirable feature which is worth more money. In the southern hemisphere it clearly has to be a northern aspect if you want energy to come into your house.

Too much solar energy entering a building can create a problem by making the interior unbearably hot and making it necessary to get rid of the excess heat. Openings for windows or doors allow a breeze through the interior to cool the inhabitants. The architecture of the hot regions of the world not only uses this method but relies on the radiative control of solar energy in its simplest form, for example by the use of highly reflective colours to help reduce solar gains during the day whilst at night the dissipation of the stored heat of the building by radiation to the very low temperatures of the cloudless sky.

Both in tropical climates and in cold northern climates windows represent a small percentage of the outside surface area of the building. It is only in temperate zones that windows have become a large percentage of outside wall surface area. The California style architecture owes its subconscious origins to climate first and energy abundance second. In the northern regions of Norway such as in Finmark there are legends about trolls who enter houses from the north side which in turn influenced stave-church architecture which avoids openings on the northern walls. One of the oldest wooden buildings in England, dating from the ninth century, is a church in Greensted Juxta Ongar, Essex, which has no windows or doors on the north side although it obviously would have benefited from them as far as lighting is concerned. It carried the Norse tradition of architecture of its time, not so much for fear of trolls as to reduce heat losses from areas through which little solar heat gain is possible.

2.1 Principles of Passive Solar Architecture

The conceptual approach to passive design considers the building as a solar collector, heat store and heat delivery system all in one, and is achieved for the most part by the sensible use of traditional architectural principles and construction techniques.

The passive utilisation of solar energy requires little or no mechanical equipment or machinery additional to the basic fabric of the building. It relies on special design solutions to maximise solar gain, mostly for space heating purposes. In order to improve heat control, passive solar energy utilisation does need simple non-energy consuming devices such as louvers and shades to direct hot and cold air through the building and to control solar radiation. In some cases fan-forced air flow is used to improve heat transfer, but this places such buildings outside the category of pure passive solar architecture even though the collection and storage may be done using the traditional parts of the building fabric.

A basic principle of passive solar architecture is the correct use of orientation so that buildings can maximise or minimise solar gains to achieve the correct comfort level inside the building. Passive solar houses often contain large areas of glass on the south facing side (in the Northern hemisphere) which may be windows to provide direct gain or glass envelopes such as conservatories. Glass transmits solar radiation and prevents cooling by the wind, hence lean-to and similar glass enclosures make walls warmer and the living space more pleasant.

One of the most important and yet simplest techniques available to the passive solar architect is the use of shading devices to control the incoming solar radiation. The Pueblo Indians in Arizona built Montezuma Castle, Figure 2.1.1, about 800 years ago. They made good use of this control principle by recessing the cave entrance to protect themselves from the intense heat of the sun in the summer when it is at high altitudes but allowing the winter sun at low altitude to warm the dwelling. Olgyay (1957) and Olgyay have summarised such methods of shading control in a comprehensive book. Kaszta (1973) and Kaszta used mechanically rotated horizontal louvers in structures overhanging the windows and recent publications of all types devote attention to this topic (for example Michels 1979). The use of strategically placed deciduous trees can achieve the same shading in the summer while admitting solar radiation in the winter after the leaves have fallen and when the sun is low over the horizon.

Sometimes reflective surfaces are employed to get more solar radiation into the house during the day or more radiative cooling at night. Since the efficiency of a reflector will depend on the angle it presents to the solar rays, tracking devices are sometimes used to optimise their performance.

Figure 2.1.1: Montezuma Castle. Indian dwellings in Arizona are recessed into cave to provide shade control.

A simple and effective solution is provided by a movable insulating shutter panel which can be set to the proper angle either to reflect solar radiation into the house during winter days or to cool the house by radiation to the sky during summer nights. As an insulator it can be shut during hot summer days and cool winter nights to protect the house from excessive temperature swings.

Another principle of passive solar heating and cooling is the use of storage. This has to be designed so that it matches the characteristic needs of energy supply and demand, as they seldom coincide. The solutions to this design problem are outlined later in this chapter.

Finally, the principle of good insulation is important for the success of passive use of solar energy. The strategic positioning of insulating layers can make a lot of difference in the use of energy. North walls have to be insulated differently from south walls and whether the insulation is on the outside or on the inside of walls will give a control over the energy flow.

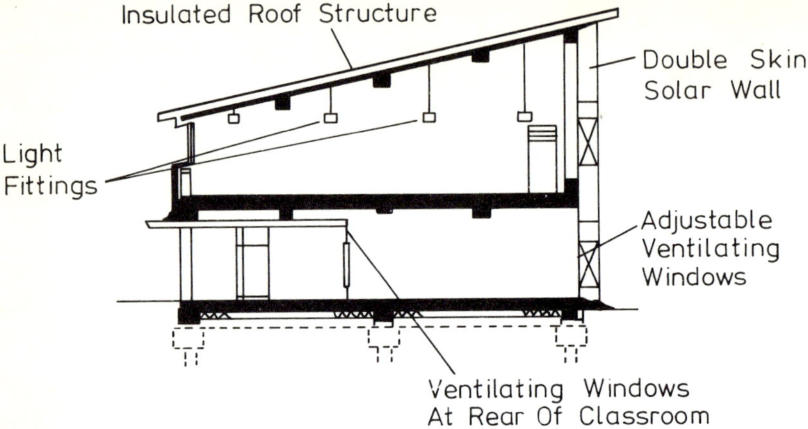

Figure 2.1.2: Wallasey school. Glass wall on the right admits solar heat.

The problem of air movement in passive solar houses either for transferring energy or to provide oxygen is occupying the thoughts of many designers. Some American designs concentrate on air movements as evidenced for instance in the survey of Anderson (1977). The case of St. George's school in Wallasey, see Figure 2.1.2, demonstrates that it is not only a simple problem of natural convection of heated air that has to be solved. Oxygen levels and body odours, cooking and combustion related problems all have to be solved. When user measures are taken (by opening windows) they sometimes experience even more odour problems or a significant loss of heat.

Page (1980) in his lecture to the International Summer School on Solar Energy, held in Canada, reviewed the principles involved in passive solar design. A structured approach to design philosophy has started to emerge and an example of that is embodied in the book written by Lebens (1980) a British architect who worked earlier at the Massachussetts Institute of Technology. It is full of useful data, worked examples and details of a passive simulation programme written for a programmable calculator.

Information on UK passive solar activity can be found in a report written for the Energy Technology Support Unit of the UK Department of Energy by British architect Turrent (1980) and his partners. It lists and evaluates a number of current UK projects and design options.

2.2 **Passive Heating**

One of the most significant conscious contributions to passive solar energy utilisation has been made by Trombe (1979) and his team at the

Solar Energy Laboratory in Odeillo, France. They have established the concept of combining the three objective principles of collection, storage and delivery into a single structure called a solar wall or a Trombe wall after Professor Trombe. It consists of a concrete wall of high thermal mass protected on the outside with a glass skin. Solar radiation penetrates the glass cover and heats the wall, which in turn heats the room on the other side by low level radiation. The space between the glass skin and the wall also acts as a duct which facilitates the natural circulation of air from the floor level of the room to the ceiling level of the room. During the passage between the wall and the glass cover the upward convecting air is heated, mainly by the wall. Figure 2.2.1 shows a diagram of the cross section of the first Odeillo house using a Trombe wall.

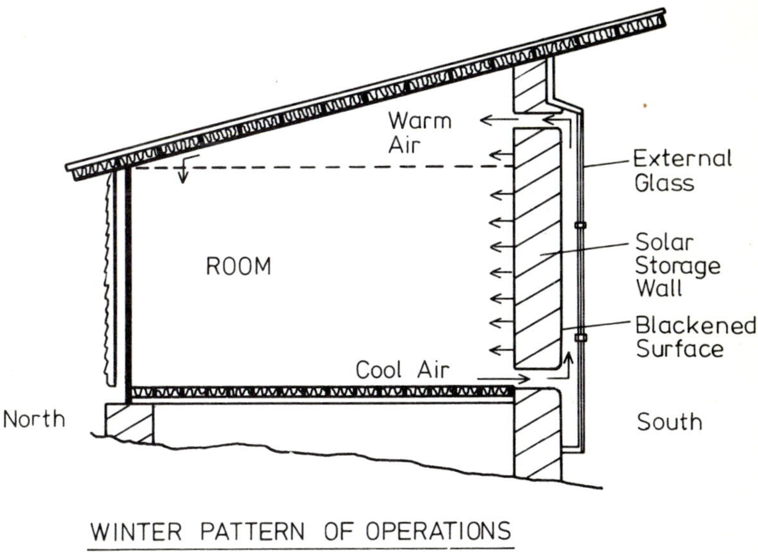

Figure 2.2.1: The first Odeillo house. In winter the heated air is in natural circulation.

Careful monitoring of the performance of the Trombe wall at Odeillo over many years revealed that the thermal conduction through the wall and hence its performance can change during the first few years due to the change in moisture content. The work at the CNRS Solar Energy Laboratory on heavy masonry solar collectors is continuing.

A similar idea has been used in the UK as the basis of a design of passive solar houses for elderly people. Greenwood (1979) and Ward started in the early seventies to work on a group of 14 houses to be built at Bebington, Wirral. These are situated in an area with a medium to severe driving rain index and where the yearly sunshine hours averaged 1451 for the past hundred years, which compares favourably with Odeillo's 2100 hours when the northern latitude and altitude above sea level are taken into consideration. The design for 9 of the houses was based on a solar wall and high levels of insulation. The solar houses were also equipped with ducts and simple control devices to help transfer the collected solar energy to the back of the house where the bedroom is situated. Figure 2.2.2 shows the cross section of a solar house in winter and summer operating modes. The remaining 5 houses were built to the same plan form but without special features and therefore could be used as a comparison. Justin (1980) and co-workers reported results following two years of monitoring both thermal performance and occupant satisfaction. This work is providing valuable data for programme validation.

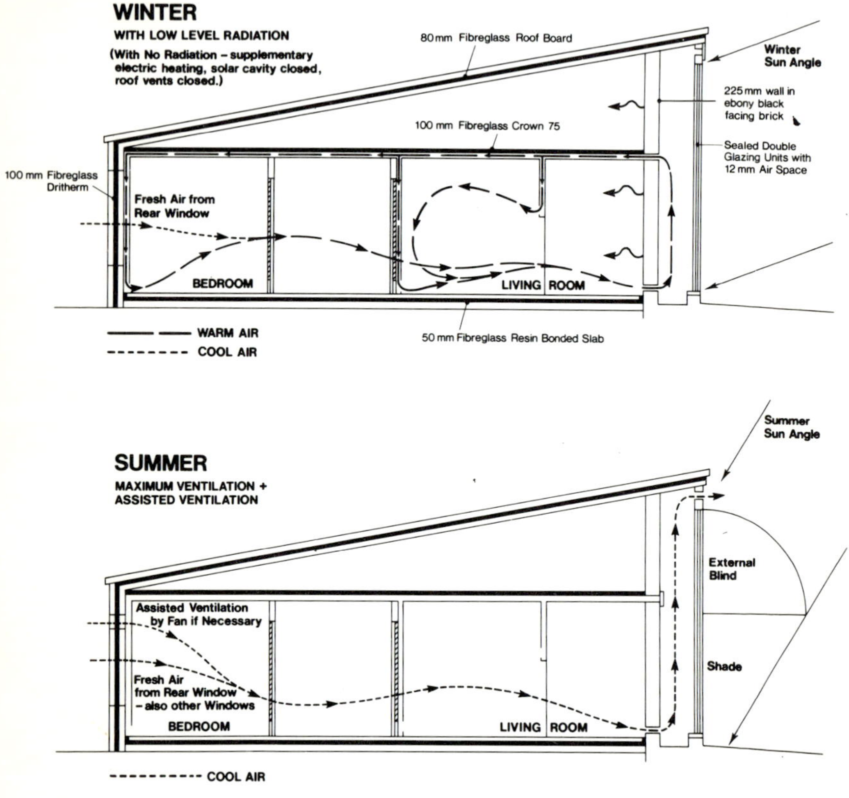

Figure 2.2.2: Acorn Close—Solar Heating and Ventillating Systems. Winter and Summer operations.

Measured performance from an outdoor laboratory testing of a solar wall was used by Lee (1980) and Fitzgerald for developing and evaluating simulation models.

Basically similar concepts as used in the solar walls are used by those who advocate solar skins for walls. McGregor (1979) described a solar skin for solid walls which consists of a single glass layer mounted 25 mm away from a 220 mm thick brick wall. In experiments at Napier College, Edinburgh, he found that the walls were warmer and drier than before applying the solar skin, the heating load was reduced and dampness problems were alleviated. Such methods work best in a colder climate and on walls with low initial thermal resistance. The protection against driving rain is also important and the economics are claimed to favour solar skins in comparison with solar water heating systems. Aesthetic considerations are the strongest argument against retrofitting solar skins but they can be improved in new designs.

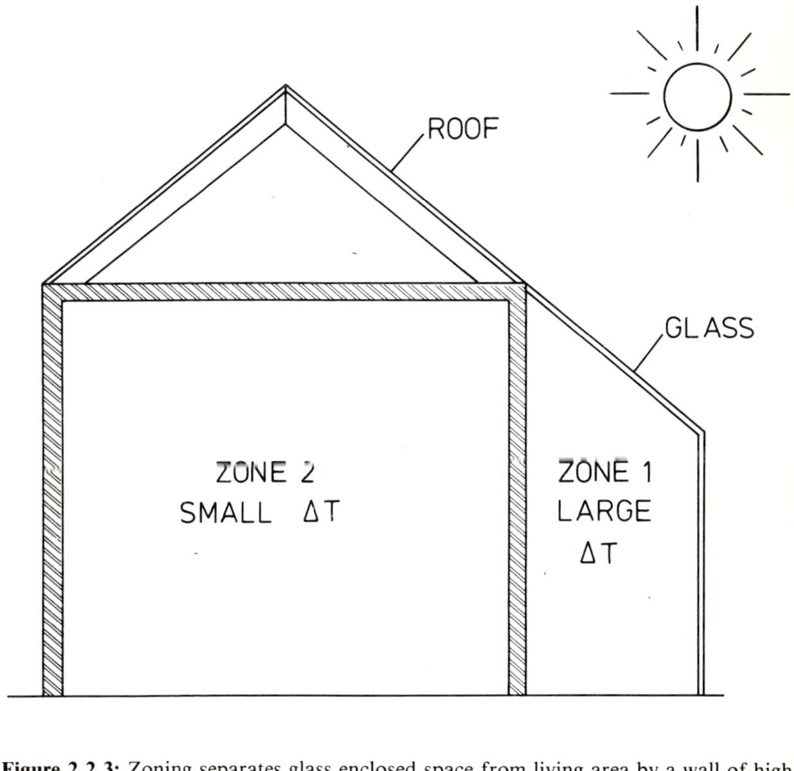

Figure 2.2.3: Zoning separates glass enclosed space from living area by a wall of high thermal inertia.

The concept of zoning which is important in passive design goes beyond the solutions offered by the Trombe wall or solar skin. It involves the use of two distinct zones in a building which are separated by a wall of considerable mass, Figure 2.2.3. Instead of glazing or skin-cladding a wall a large space between the glass enclosure and the wall is created (zone 1) which enables certain living functions of the house to be carried out in this area, but with the limitation that the temperature swings are wider in this zone. Conservatories, lean-to structures and sun terraces enclosed in glass are examples of this zone. Balcomb (1979 A) working at the Los Alamos Scientific Laboratory in New Mexico recognised the importance of control for these spaces in order to reduce the excessive temperature swings, which can reach 30-35°C during clear winter sky conditions. The space behind the wall (zone 2) is where the temperature swings are narrower and all living functions of the house are possible. The excess heat from zone 1 is transferred to the wall directly and to zone 2 by natural air circulation. This reduces the swings in zone 1 and keeps the buffered area of zone 2 at a higher and more constant temperature by virtue of using it as an additional storage mass.

An example of this type of construction is the house Superhome 2000 at Milton Keynes which featured in a special series of programmes of BBC2's 'Money Programme'. The house which is insulated to approximately twice the level required by building regulations, contains on its south side a double-height conservatory. Heat from the conservatory can be drawn into the main living volume with the aid of small fans.

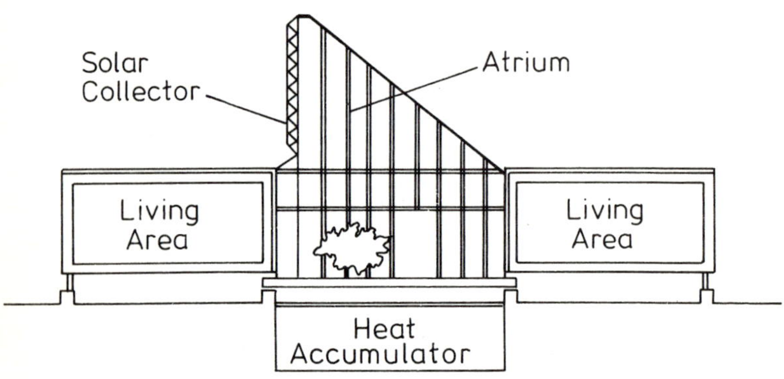

Figure 2.2.4: The Danish "Zero Energy House". South is to the left.

A buffer zone in the middle of a solar house was used in the Danish 'Zero' Energy House completed in 1975, shown schematically in Figure 2.2.4, although for the most part it is an active solar system. A somewhat

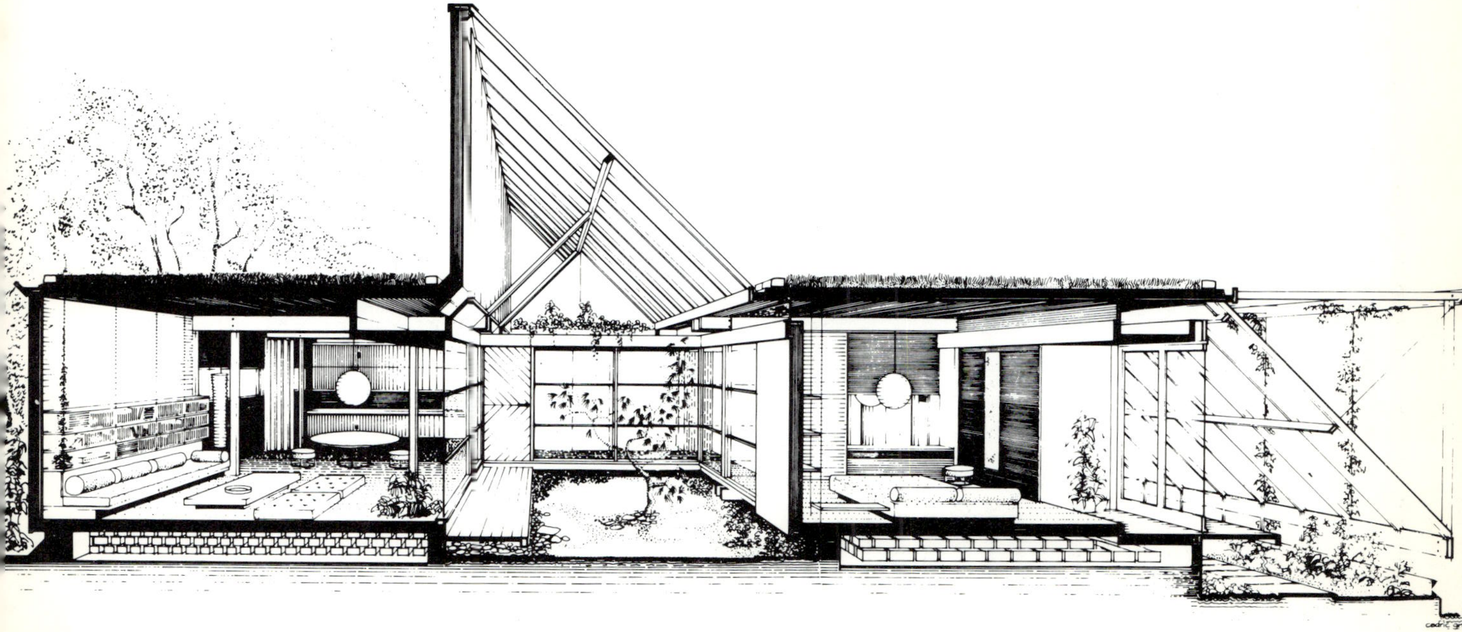

Figure 2.2.5: Green's "Shed". South is to the right.

similar arrangement of a central glass covered atrium is proposed by Green (1979) in his Shed house shown in Figure 2.2.5. The important difference, however, is that his Shed with its central court has south facing sloping surfaces whereas in the Danish house these face north. We can see in this drawing an interesting combination of glazed conservatories followed by insulated living spaces as one progresses from south to north through the building. Green's design won third prize in a Japanese competition in 1978, and second prize in the CEC competition in 1980.

Successful passive solar houses like the one designed by Green make use of direct gain through large window area, controlled by shutters and by summer shading. They also use insulation to a much greater extent than conventional architecture did until recently.

It would be a mistake to think that the obligatory orientation of passive solar houses renders them to boring monotony in their outward appearance in large housing projects. The Pennyland passive solar housing project was described recently by Doggart (1980) and Everett. In this development there are 94 houses with high standard of insulation and 83 houses with very high standard of insulation. They are arranged by the London Architects Energy Conscious Design Limited, in an imaginative pattern which does not give the impression of monotony, yet allows passive solar energy utilisation through direct gain with little loss due to misalignment in orientation. The performance of the two groups is being monitored with funding from the UK Department of Energy.

The role played by windows and other glazed openings were investigated by Davies (1980) who used 50 years of weather data to analyse windows and glazed walls on a building located at 53.4°N. He concludes that in the winter they can be thermally disadvantageous unless special measures are taken. Page (1981 B) surveyed the various methods used in Northern Europe in reducing heat loss through windows at night. He emphasised the various aspects of comfort, other than temperature, which have to be considered in passive solar design

A notable pragmatism characterised the recent German passive solar design competition and demonstration project which produced successful co-operative efforts like the one between the Berlin Institute of Building Environment and Solar Technology represented by Schreck (1980) who contributed the energy concept and Ungers who was in charge of the design. They have considered every aspect of design from urban planning to heat storage. In Britain there are several active groups of architects, engineers and designers who spearhead the new approach to building design.

The first European Passive Solar Competition, funded by the CEC and supported by the Union Internationale des Architects in Paris, was organised by Ralph Lebens Associates in London. In the single house group Perraudin from France received the first prize and Raeck from Germany the second. English, Italian, Scottish and Belgian entries were also winners. In the clustered housing category the first prize went to the Danish trio: Lundgaard, Rotne and Sorensen. The second prize was jointly awarded to Green from England and Barrett from Ireland. The other winners were from France, England and Italy. In the multi-storey section no first prize was awarded. The second prize went to a British project which retrofitted a stone terraced building in Bath. This 1980 event was the first of a series of such competitions planned by the Commission.

2.3 **Hybrid Passive and Active Heating**

The arguments in favour and against mixed or hybrid solar designs using both passive and active solar energy conversion was studied by Balcomb (1979 B) and co-workers mostly by simulation using the program LASL SOLAR developed at Los Alamos Scientific Laboratories. At about the same time Kiraly (1980) in Austria built a hybrid active-passive solar house of 150 m^2 living floor area (excluding the swimming pool and the service area). His well insulated single family dwelling uses a conservatory and a high density masonry store as its passive elements and combines these with an active system which includes 30 m^2 of flat plate solar collectors and a heat pump connected to an underground heat exchanger. These feed an underfloor heating system and an indoor swimming pool which also serves as a heat store. According to the designer-builder-occupant the passive features reduce the heating load to 30 MWh of which 22% is supplied by the solar collectors, 49% comes from the soil and 29% from electricity. This example shows that hybrid systems can cope with weather conditions of the Inn Valley (47°15 N, 950 m above sea level).

Another interesting hybrid system is one designed by Wells for a house in Raven Rocks, Ohio, and described by Anderson (1977). The house is partly buried under the soil on the northern side. Large southern sloping glass walls admit solar radiation into the building where massive concrete structures absorb it. There is a rock bed under the lower floor and air can be blown through that for later use. Insulating panels cover the windows at night and the sloping front accelerates the wind which is channelled through convergent-divergent nozzles where wind turbines convert the energy.

Nearer home, the Ideal Home Solar House at Milton Keynes has high levels of insulation and a two-storey south facing conservatory to minimise the space heating requirements. In addition the house is equipped with 27 m^2 of flat plate collectors coupled to a Patent heat store. Dominic Michaelis, the noted British architect, designed this house. He was also successful in a number of passive solar house designs in various European competitions.

2.4 **Passive Cooling**

There are many areas of the world where the main concern in building design is how to get rid of the excess solar energy entering the building. In all cases the best policy is to prevent the solar energy from entering but sometimes this is not possible for various reasons. Shading by trees and man made shades will certainly be simple and cost effective measures. For better control over the cooling of a building it is necessary to use either convective or radiative methods, or a combination of both.

The convective methods usually rely on reversed air flow at night, such as the method used in the Odeillo houses, Trombe (1979) and co-workers, or in the Davis house in Albuquerque, New Mexico, described in many publications such as Lebens (1980). These convective methods require an outside air temperature to be lower than the temperature of the structural fabric of the house. The convection can be enhanced by using wind towers such as the ones built in Iran. Bahadori (1978) describes the main air flow patterns prevailing during windy and windless nights. He claims that 5-10°C sensible cooling can be achieved.

Work done in the UK on passive ventilation alternatives was reported by Dodson (1979) who considered the case of passive solar cooling by ventilation and gave some practical formulae for calculating flow velocities. Using the chimney effect he discussed passive cooling of houses with and without a buffer zone like a conservatory.

Radiative cooling relies on the effective sky temperature being lower than the outside surface temperature of the house. Cloudy and cloudless nights will have different sky temperatures, in very hot arid zones the sky temperature at night is usually low due to the lack of clouds. One interesting example of radiative cooling is Steve Baer's house in Corrales, Alburquerque, New Mexico. As the sketch in Figure 2.4.1 shows the heat store consists of oil drums filled with water and painted black on the outside end while they are painted light cream colour on the end which faces the room. A moveable outside panel, hinged at the bottom, can be manipulated by pulleys and steel cables fastened to a crank inside the house. The panel acts as an insulator when it is closed. Since the inside surface is coated it can also act as a mirror. During the night this moveable mirror can be lowered into a position which would reflect the radiant heat coming from the blackened ends of the drums to the sky. This is an effective way of enhancing radiative cooling as the drums will always 'see' the low temperature sky. As an added bonus the reflective panels can also enhance the radiative gains during the day if that is desired. They are simple, cost effective and their control is obvious; you crank the cable until you obtain the maximum heat or coolness.

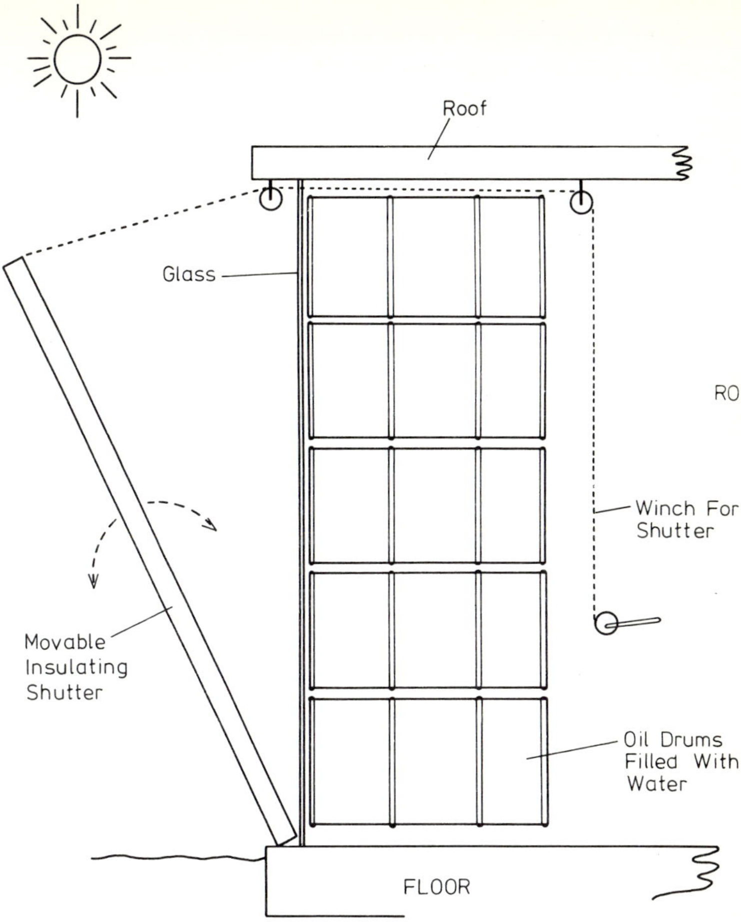

Figure 2.4.1: Steve Baer's house in New Mexico. The movable shutter acts as a reflector, when open, for cooling at night and heating in daytime.

Another effective way to cool a passive house is by the use of roof ponds. A system based on water contained in plastic bags placed on a flat roof top was first designed by Hay (1979). Moveable insulating panels above and below the pond can control the direction of heat flow by radiation. It is like having a Trombe wall on your roof which you can allow to radiate downwards at night for heating or upwards for cooling. This control aspect makes the application a novelty. Air-conditioned houses have used the technique of flooding flat roofs for lowering heat loads and reducing temperature swings for many decades in Europe and the North American continent.

2.5 **Greenhouses**

Plants grow better, healthier and faster if we protect them from large temperature swings. Greenhouses work by letting the solar radiation through the glass and at the same time providing an effective protection against convective cooling by wind. They have been used extensively north of 45° northern latitude. More recently, however, there has been an increased interest in greenhouses in the tropical zone and this has helped to focus attention on two problems associated with greenhouses: the need for heating and the need for cooling.

In the UK for instance the difference between summer and winter radiation levels is so great that any greenhouse left to its own devices will not be able to maintain a narrow enough temperature range for long periods of time. The traditional solution in the winter was to heat the greenhouse with oil burning heaters which would also produce CO_2 as a welcome by-product. During the summer the glass could be painted white to reduce the incoming radiation and windows could be opened. The need for these measures reflect the poor design of traditional greenhouses. As the demand for greenhouses is now from about 30 to 65° northern latitude, there is an urgent need to find the solution to cooling, heating, energy storage and optical control of solar radiation in greenhouses.

Different designs suited to particular climatic conditions have been prepared to improve thermal performance. Figure 2.5.1 illustrates some designs as reviewed by Sayigh (1979).

In India Mannan (1979) and Cheema studied the cooling and heating of greenhouses at 31°N. They used orientation and geometry as one of the controls and a thin layer of water on the roof as another. Water helped to reduce the greenhouse temperatures by 10-12°C during sunshine hours. Their monitoring showed that in winter about 5°C temperature difference was maintained during the night, while twice that was observed during the day. These winter temperature differences, relative to the outside ambient temperature, were achieved without energy storage.

Passive heat storage and natural ventilation is incorporated in the design of a commercial greenhouse in Tennessee (36°N). McGowan (1979) based his design on computer simulations and used water filled drums on the north side of the greenhouse as a heat store. Air enters on three sides and exits through the apex of the roof. The average temperature difference from November to March was 10.7°C. In the summer the greenhouse temperatures were 4.5°C lower than outside air temperatures in shade. The fibreglass reinforced acrylic glazing gave a diffuse light inside.

Recently Morrison (1980) reported on experimental and analytical comparison of test results obtained in Sydney, Australia (34°S). He studied the collection of energy from the moist greenhouse atmosphere and storage in the ground by thermosyphone circulation. This mechanism was found to

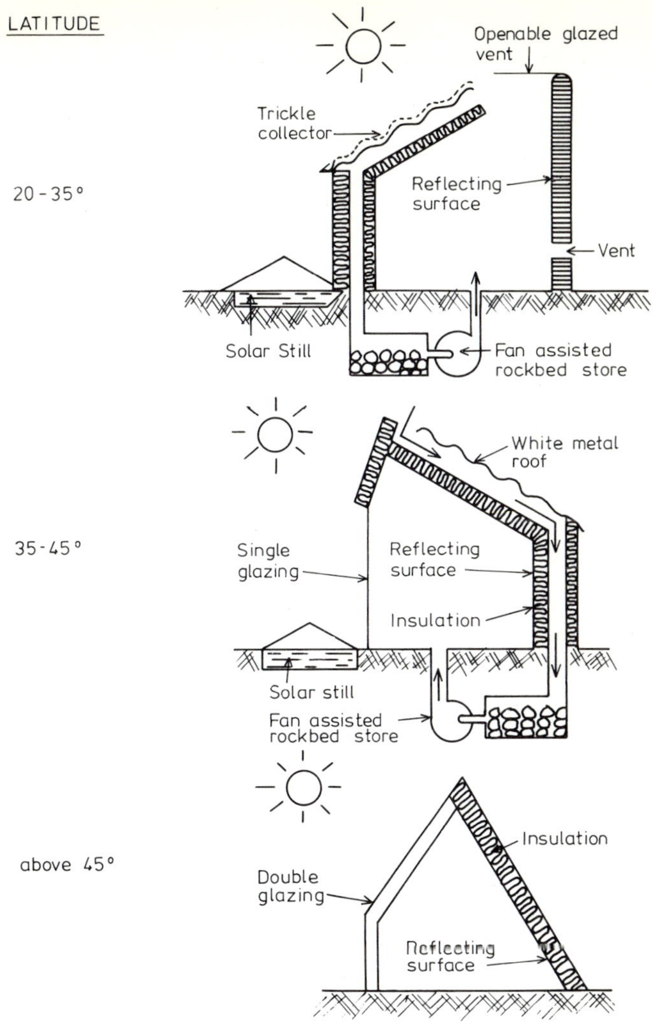

Figure 2.5.1:

be affected by cloud conditions which could reduce the circulation velocity from 14 cm/sec to less than 3 cm/sec and thus a deterioration in the heat transfer followed.

Greenhouses in hot climates, like Saudi Arabia which lies 20-30°N were studied by Sayigh (1979). He reported on experimental projects and future designs, some of them very imaginative. In one case copper chloride is pumped in a double skinned roof of the greenhouse which helps to control absorption of infrared radiation and acts as a heat exchanger medium giving its heat to desalination or crop drying processes. These, however, are on the fringe of passive solar energy utilisation as they require much active intervention.

The use of copper chloride as a selective filter was first proposed by Damagnez (1976) as a means of preventing radiative cooling at night and over-heating during the day in a Mediterranean climate (41°N). An experimental greenhouse using a copper chloride selective filter operated at the Desert Research Institute in Israel is shown in Figure 2.5.2. For higher latitudes (60°N) the Brace Research Institute has developed several designs as reported by Lawand (1979) and co-workers.

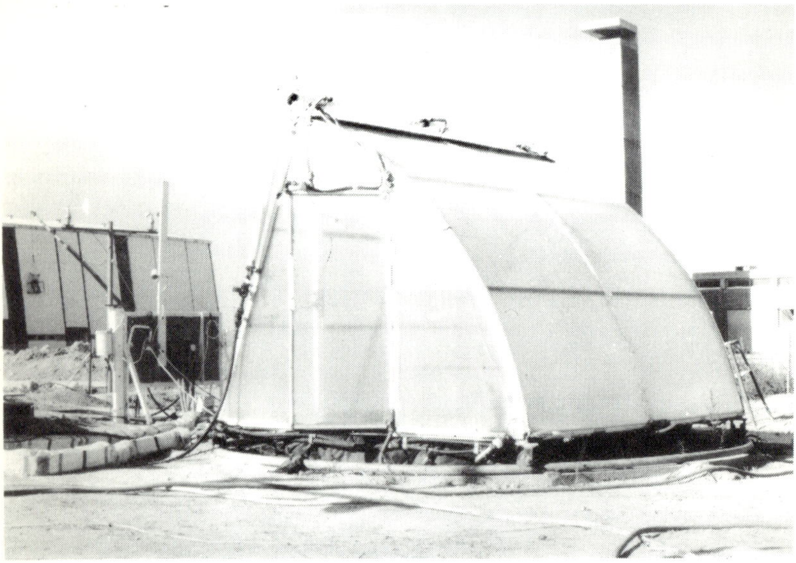

Figure 2.5.2

2.6 Special considerations in passive systems

This chapter on passive solar energy, whether in buildings or in greenhouses, has shown the overriding concern of designers and operators about the temperature changes and temporal variations in passive applications. It must not be forgotten that there is more to comfort and proper living conditions, whether for people or for plants, than temperature. Humidity, air quality and air movement are also important. The ASHRAE (1977) Handbook of Fundamentals and the CIBS Guidebook have chapters with much practical information on the subject. The best monograph was written by Fanger (1972) who discusses comfort in the context of environment, clothing and human activity.

As passive solar heating became more of a conscious design objective there was a clear need to update the work of Fanger. Wray (1980) working at Los Alamos developed a method to linearise the Fanger thermal comfort equation, introduced an equivalent uniform temperature which helps assessing thermal comfort levels in passive solar heated buildings and non-uniform environments.

Heat storage is a very important part of the passive solar design. In the demonstration building of the University of Delaware, called Solar One, a chemical material which undergoes a phase change was used in the front elevation of the house to store passive solar gains. The storage system was designed by Telkes (1974) but apparently it could not live up to expectations and Telkes embarked on a program to develop a series of chemical agents to gain better control over two way phase change which has to be repeated several thousands of times for it to be economical and practical.

A difficulty arises if a floor is intended to be used as a storage mass since it cannot be covered with rugs or carpets due to their insulating effect. This could be overcome if the heat is stored in the ceiling and radiated from there. One of the experimental houses at the Massachusetts Institute of Technology uses window blinds to reflect the incoming radiation onto a storage system built into the ceiling. The ceiling tiles contain sodium sulphate decahydrate which has the right phase change temperature and can store large amounts of energy. Some of the problems associated with providing proper adhesion of the heavy tiles to the ceiling has been plaguing the project but several solutions are being worked on.

The Institute of Energy Conversion at the University of Delaware reported the successful application of a modular element the Thermal Wall Panel which uses $NA_2SO_4.10H_2O$ in plastic tubes. Faunce (1979) and co-workers claim a 7-10 years payback period for such systems.

2.7 **Monitoring Passive Buildings**

Passive solar buildings require somewhat different considerations than active systems when designing experiments to monitor their performance. Material properties have to be measured as described by Trombe (1979) and co-workers, natural air currents have to be investigated as reported by Justin (1980) and co-workers. The instrumentation may be similar as far as the data logger is concerned to that used in active system monitoring. Lebens (1980) gives comparisons with cost estimates of two monitoring set-ups. The Energy Technology Support Unit commissioned a survey of data logging equipment available in Britain in 1978 and came out in favour of the Microdata 1600 system; Mitchel (1978). In some cases, however, people develop their own logging and control systems. Fowlkes (1979) reported the development, installation and successful use of an inexpensive data acquisition system based on a readily available Radio Schack (Tandy) computer, the TRS-80. He developed a battery operated power supply and an analogue-digital interface. The system performed 700 scans per hour and printed out average values. The ordinary Philips cassette used in high fidelity music systems held 11,000 numbers on each side.

Economic considerations frequently lead to retrofitting solar installations or solar remodelling of older buildings. In London there are many properties which can benefit from such retrofit applications. Turrent (1981) and his partners at Energy Conscious Design have designed a combination of passive and hybrid elements with controlled ventilation and air heat recovery for ten existing 4 storey Victorian terraced houses. They predict a solar contribution of 52%.

Mickelson (1979) and Sizemore described a project carried out on an Atlanta, Georgia, school building. The 50 year old building with trees around it turned out to be very energy efficient.

The economics of passive solar houses in developing countries represent a special case. Tedeschi (1978) and co-workers consider social and technical aspects and recommend a figure of merit to be used which implies a payback period of 5 or more years on the extra cost due to passive solar application. They warn about the importance of estimating the economic consequences of storage correctly which is more difficult to do in passive than in active solar systems.

In Germany Steinmuller (1980) studied the thermal effects of solar building design for Central European weather conditions. Working at the Philips Research Laboratories at Aachen he found that considerable passive solar gains are possible in poorly insulated houses, while the heating load is only just reduced in buildings of high insulation standards with controlled ventilation. His work on the optical and thermal properties of windows has been carried out in the same laboratory with Horster and others. Bruno (1979) and co-workers indicate the concern felt by the Philips group for passive solar energy applications.

In May 1980 the US Department of Energy organised an Internal Expert Group Meeting on passive and low energy cooling, heating and dehumidification at the University of Miami in Florida. Page (1981 A) and also others brought attention to the many aspects of passive design. Better understanding of climate, building materials, construction methods, structural elements, user demands, heat transfer in unusual geometries, fluid flow in closed spaces and optical characteristics of glazing materials is needed: Sharples (1980) and Page, Hollands (1979).

Selectively behaving glasses are being developed in several countries which can make important differences, just as selective coatings on collector surface brought in new ways of thinking a few years ago.

The basis of passive solar energy utilisation is good architecture which uses simple technologies and high levels of insulation. Successful designs can combine these into solutions which provide buildings with good thermal comfort and low energy demand at reasonable cost, whilst remaining aesthetically pleasing. The popularity of passive solar heating and cooling is likely to increase in the coming years. Some architects would say that they have been in the passive solar energy business all their life, before the current fashion discovered it and gave it a new vocabulary.

References

Anderson, B. (1977): **Solar Energy: Fundamentals in Building Design.** McGraw-Hill 1977, 374 p.

ASHRAE (1977): Handbook and Product Directory, 1977, Fundamentals American Society of Heating, Refrigerating and Airconditioning Engineers, New York, 1977.

Bahadori, M. N. (1978): "Solar energy utilisation for developing countries." in **Proc. 1st International Conference on Solar Building Technology,** London 25-29 July 1977. ed. C. Stambolis. RIBA Publications Ltd. 1978, Vol. 2, 522-530.

Balcomb, J. D. (1979 A): "Passive solar heating for buildings." in **The Passive Collection of Solar Energy in Buildings.** Conference (C19) London, UK-ISES 1979, 75-85.

Balcomb, J. D. (1979 B), J. C. Hedstrom and R. D. McFarland: "Hybrid active-passive designs; synergy or conflict?" Proc. ISES Silver Jubilee Congress in Atlanta 1979, Vol. 2, 1556-1560.

Bruno, R. (1979), H. Hörster, H. Köstlin, G. Soellner and B. Steinmüller: "Optical and thermal properties of windows for passive solar systems." Proc. ISES Silver Jubilee Congress in Atlanta 1979, Vol. 2, 1651.

CIBS Guide Book: Chartered Institution of Building Services Part A: Design Data, Part B: Installation and equipment data, Part C: Reference data. Available in separate sections, revised from time to time since 1974. Published by CIBS, London.

Damagnez, J. (1976): "Need and design for a solar greenhouse." in **Solar Energy in Agriculture.** Conference (C9) London, UK-ISES 1976, 82.

Davies, M. G. (1980): "Useful solar energy gains through glazed and opaque structures." in **Solar Energy in the 80s.** Conference (M2) Birmingham, UK-ISES Midlands Branch 1980, 32-37.

Dodson, C. (1979): "Passive ventilation alternative." in **The Passive Collection of Solar Energy in Buildings.** Conference (C19) London, UK-ISES 1979, 33-38.

Doggart, J. (1980) and R. Everett: "Passive solar housing at Milton Keynes." in **European Community Solar Housing.** Conference (C23) London, UK-ISES 1980, 40.

Fanger, P. O. (1972): **Thermal Comfort Analysis and Application in Environmental Engineering.** McGraw-Hill 1972.

Faunce, S. (1979), J. Sliwkowski and S. Guceri: "An application of phase change materials in passive solar systems." Proc. ISES Silver Jubilee Congress in Atlanta 1979, Vol. 2, 1637-1641.

Fowlkes, C. W. (1979): "Measured performance of a passive solar residence in Bozeman, Montana." Proc. ISES Silver Jubilee Congress in Atlanta 1979, Vol.2, 1600-1604.

Green, C. (1979): "The Shed – construction of a solar heated experimental dwelling in Sheffield." in **The Passive Collection of Solar Energy in Buildings.** Conference (C19) london, UK-ISES, 55-59.

Greenwood, P. (1979) and H. Ward: "Solar houses for the elderly, Acorn House, Wirral. A development by Merseyside Improved Houses." in **The Passive Collection of Solar Energy in Buildings.** Conference (C19) London, UK-ISES 1979, 60-74.

Hay, H. R. (1979): "Thermopond applicability to climate and structure." **Proc. ISES Silver Jubilee Congress in Atlanta 1979,** Vol. **1,** 1535.

Hollands, K. G. T. (1979): "Free convection in solar collectors." in **Solar Energy Conversion.** (5th Course, Univ. of Waterloo, Canada, 1978) ed. A. E. Dixon and J. D. Leslie, Pergamon Press 1979, 126-147.

Justin, B. (1980), A. G. Guy and G. Shaw: "Monitoring solar walls in occupied houses." in **Modelling & Monitoring Solat Thermal Systems.** Conference (C24) London, UK-ISES 1980, 94-104.

Kaszta, J. (1973) and D. Kaszta: Private communication.

Kiraly, J. J. (1980): "Solarhaus Kiraly. Ein passives und aktives Solarheizsystem." **Tagungsbericht des 3. Internationalen Sonnenforums - Freie Vorträge -** Hamburg 1980 DGS-Sonnenenergie Verlag, 564-572.

Lawand, T. A. (1979), J. LeNormand, A. Skelton, M. Dorgand and J. MacArthur: "Investigations with a heat storage – night cover Brace design greenhouse for colder regions." **Proc. ISES Silver Jubilee Congress,** Atlanta 1979, Vol. 2, 1713.

Lebens, R. M. (1980): **Passive Solar Heating Design.** Applied Science Publishers 1980, 234 p.

Lee, J. B. G. (1980) and D. Fitzgerald: "Comparison of physical and mathematical models of solar walls." in **Modelling & Monitoring Solar Thermal Systems.** Conference (C24) London, UK-ISES, 1980, 84-93.

McGowan, T. F. (1979): "Commercial growers passive solar greenhouse at the Hidden Springs Nursery." Proc. ISES Silver Jubilee Congress in Atlanta 1979, Vol. 2, 1704-1707.

McGregor, A. W. K. (1979): "A solar skin for solid walls." in **The Passive Collection of Solar Energy in Buildings.** Conference (C19) London, UK-ISES 1979, 39-43.

Mannan, K. D. (1979) and L. S. Cheema: "Year-round studies on natural cooling and heating of greenhouses in northern India." Proc. ISES Silver Jubilee Congress in Atlanta 1979, Vol. 2, 1699-1703.

Michels, T. (1979): **Solar Energy Utilisation.** Van Nostrand Reinhold 1979, 253 p.

Mickelson, J. L. (1979) and M. M. Sizemore: "Economic aspects of passive solar remodelling in existing buildings." Proc. ISES Silver Jubilee Congress in Atlanta 1979, Vol. 2, 1569-1573.

Mitchell, H. J. (1978): **Data Logging System Review.** Energy Technology Support Unit, Harwell 1978, 2 vols.

Morrison, G. L. (1980): "Passive energy storage in greenhouses." **Solar Energy 25** 1980, 365-372.

Olgyay, V. (1957) and A. Olgyay: **Solar Control and Shading Devices.** Princeton University Press 1957.

Page, J. K. (1980): "Passive solar building design." Paper presented to the International Summer School on **Solar Energy.** University of Ontario, London, Ontario.

Page, J. K. (1981 A): "Passive solar building design." Proc. of the International Expert Group meeting on **Passive and Low Energy Cooling, Heating and Dehumidification.** Univ. of Miami, Florida, May 1980. US Dept. of Energy, 1981.

Page, J. K. (1981 B): "Windows as solar collectors." Lecture given at the UK-ISES Midlands Branch meeting in Birmingham on 12 Feb. 1981.

Sayigh, A. A. M. (1979): "Greenhouses in hot climate." in **Solar Energy Conversion.** (5th Course, Univ. of Waterloo, Canada, 1978) ed. A. E. Dixon and J. D. Leslie, Pergamon Press 1979, 1145-1168.

Schreck, H. F. (1980) and O. M. Ungers: "Entwürfe für eine Solartypologie Wettebewersarbeit für das Demonstrationsprojekt Melkerei Landstuhl." **Tagungsbericht des 3. Internationalen Sonnenforums - Freie Vorträge -** Hamburg 1980, DGS-Sonnenenergie Verlag, 513-515.

Sharples, S. (1980) and J. K. Page: **Natural Convective Heat Transfer Across Parallel Sided Air Cavities in Building Components – A Review of Current Knowledge.** Building Science Internal Report BS51, Dept. of Bldg. Science, Univ. of Sheffield 1980.

Steinmüller, B. (1980): "Auswirkungen solararchitekt onischer Massnahmen auf die Energiebilanz von Gebäuden." **Tagungsbericht des 3. Internationalen Sonnenforums - Freie Vorträge -** Hamburg 1980, DGS-Sonnenenergie Verlag, 554-563.

Tedeschi, E. (1978), C. de Rosa, R. Vilaprino, G. Lelio and E. Fernandez: "Solar energy applications for buildings in developing countries." in **Proc. 1st International Conference on Solar Building Technology,** London 25-29 July 1977. ed. C. Stambolis. RIBA Publications Ltd. 1978, Vol. 2, 531-541.

Telkes, M. (1974): "Storage of solar heating and cooling." **ASHRAE Journal** Sept. 1974, 38-44.

Trombe, F. (1979), J. F. Robert, M. Cabanat and B. Sesolis: "Some performance characteristics of the CNRS solar house collectors." in **The Passive Collection of Solar Energy in Buildings.** Conference (C19) London, UK-ISES 1979, 4-23.

Turrent, D. (1980), J. Doggart and R. Ferraro: **Passive Solar Housing in the UK.** A Report to the Energy Technology Support Unit, Harwell. Energy Conscious Design, London 1980, 103 p.

Turrent, D. (1981): Private communication.

Wray, W. O. (1980): "A simple procedure for assessing thermal comfort in passive solar heated buildings." **Solar Energy 25** 1980, 327-333.

Chapter 3

ACTIVE COLLECTION SYSTEMS

The active collection and use of solar energy goes beyond the architectural aims of passive systems. It uses pumps, valves, motors, mechanisms and transducers in an attempt to achieve better results. The price to be paid for this is the involvement in technology that inevitably follows.

Active collection of solar energy may be used for a wide range of applications such as water heating, space heating and industrial processes. Active systems have the advantage that they remove energy from the collectors in a controlled way and the heat may be stored for use when required, thus improving overall efficiency. By an appropriate choice of collector and system design a wide range of temperatures can be achieved. Thus active systems include those which use simple flat plate collectors to heat a swimming pool, those which use concentrating collectors to generate steam, and even a solar furnace which can reach 6000 K.

3.1 Flat plate collectors

Although many of the earliest solar collectors employed concentrating mirrors and lenses, it is the flat plate solar collector which has become popular in recent years. In the UK the first scientific investigation of flat plate collectors was made by Heywood (1954) whose suggestions for the design of domestic solar water heaters are still used today.

The essential component of a flat plate collector is a planar absorber incorporating channels through which a heat transfer fluid is passed to remove the solar energy. For low temperature applications such as swimming pool heating this may be sufficient, but when higher temperatures are required it is necessary to reduce the heat losses from the absorber by enclosing it in an insulated box and glazing the side facing the sun, as shown in Figure 3.1.1. A good many books describe flat plate collectors. The UK-ISES publication "Heating Water by the Sun" edited by Gillett (1981) and Hutchins describes in simple terms the practical aspects of flat plate collectors used for heating domestic water and swimming pools. A widely used engineering text is that by Duffie (1980) and Beckman, which gives detailed analysis of performance for the various collector designs.

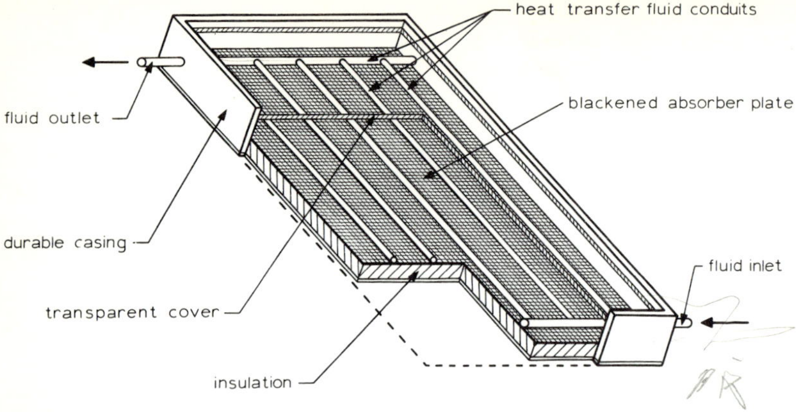

Figure 3.1.1: Cutaway view of a flat plate solar collector.

Flat plate collectors are popular for swimming pool and domestic water heating. They can be made in modules of relatively simple design and lend themselves to mass production. They are well suited for fitting to existing buildings which for the next couple of decades is likely to be important. Sloping roofs, common in the northern hemisphere, are the natural surfaces on which to mount solar collectors unobstructed to receive radiation. Good advice on how to make sure the installation will have a long life is detailed by Wozniak (1979) in a Building Research Establishment Report.

An understanding of how flat plate solar collectors work is necessary if high performance, well engineered, durable collectors are to be produced. The problems of how to maximise the absorbed energy and minimise the losses have been the subject of much research and development in recent years.

An analytical approach to collector thermal performance was first made at the Massachusetts Institute of Technology where Hottel (1942) and Woertz started to evaluate the energy balances of flat plates exposed to solar radiation. During the next seventeen years Whillier and Bliss also contributed to the expression and their names are identified with collector efficiency through the Hottel-Whillier-Bliss equation.

The performance of a collector can be described using two coefficients. One, the optical coefficient, describes the degree to which the solar radiation is absorbed and the other the extent of heat losses to the collector's environment. The heat loss to the environment increases when the temperature of the collector rises above that of its surroundings. The collector's efficiency, which is the fraction of the incoming radiation that can be usefully removed by the heat transfer fluid, decreases as the collector temperature rises. This is shown in Figure 3.1.2 for different types of flat plate collector. For any collector the slope of the curve is proportional to its heat loss coefficient and the intercept with the efficiency axis relates to the degree of absorption of solar radiation. Thus an unglazed collector absorbs more of the incoming radiation than does a glazed collector but cannot reach as high a temperature because of its greater heat losses.

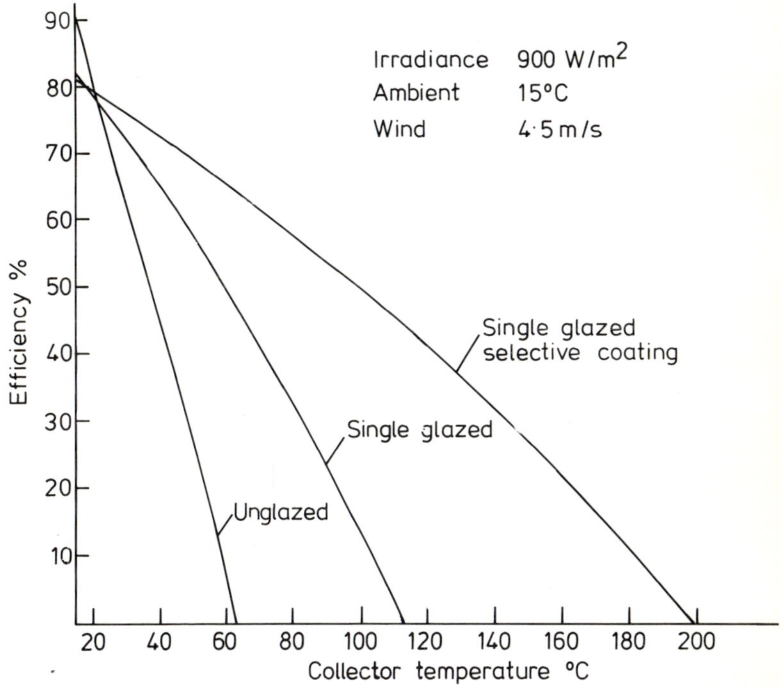

Figure 3.1.2: Collector efficiency versus plate temperature. Typical operating range is 40° - 100° C.

The optical coefficient depends on the transmission properties of the glazing and on the absorptance of the surface coating on the absorber plate. Kenna (1981) has studied the optical properties of a range of collector cover materials including both plastics and glasses. The treatment of glasses with anti-reflective coatings is also receiving attention, for example Mar (1975) and co-workers reported on the use of optical coatings for flat plate collectors.

Apart from conduction losses through the insulation behind the absorber plate heat is also lost through the front of the collector by convection and radiation. Workers at the University of Waterloo have studied convective heat losses from collectors. Hollands (1976, 1979) and co-workers have derived expressions for convective heat transfer across inclined air layers which can be used to select the spacing between cover and absorber. Hollands (1965) proposed the use of honeycombs placed between the absorber and the cover plate to suppress convection, in his later paper Hollands (1978) and co-workers discussed the solar transmittance of such devices. Shewen (1979) and Hollands examined the factors affecting the heat loss coefficient and the influence of this on collector performance.

Heat losses by radiation from the absorber to the cover can be reduced by the use of selective coatings applied to the absorber or the cover. These are discussed in Section 3.3.

The efficiency of the collector is affected by the ease with which the energy absorbed by the plate is transferred to the fluid. This in turn varies with the design of the absorber plate and as discussed by Jesch (1981 A) and co-workers with the flow rate of the heat transfer fluid.

Kenna (1981) has shown that the ratio of the heat loss coefficient to the optical coefficient is an important parameter which indicates the ability of the collector to provide high grade energy and is therefore useful in comparing collectors.

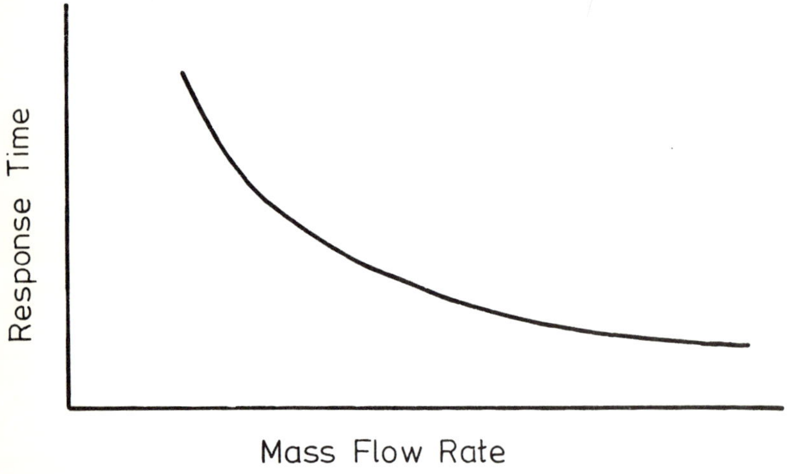

Figure 3.1.3: Response time versus mass flow rate in flat plate solar collectors exposed to sudden change of irradiance.

Since the solar irradiance to which the collector is exposed can change rapidly another important parameter affecting the collector's performance is its thermal inertia. Klein (1974) and co-workers treated this problem analytically and Rogers (1979) proposed a model to generalise the Hottel-Whillier-Bliss equation for transient behaviour. He used a linear response function to relate the useful output power at each time to the net power absorbed by the plate during the previous time. Jesch (1981 A) and co-workers have measured the effect of flow rate on thermal response time on differently designed and constructed collectors. As can be seen in Figure 3.1.3 the response time to a sudden change in irradiance reduces as the flow rate increases until an appropriate value is reached. They also found that although the thermal mass of the plate is a prime factor, the optical coefficient and the heat loss coefficient affect the response time.

The testing of collectors has become a central issue in many countries. The need to compare relatively unknown products has put pressure on the solar industry to regulate itself and in many countries the industry, Governments and standards organisations have issued codes of practice and eventually regulatory standards. Test methods which have been widely adopted include a steady state outdoor test which originated at the National Bureau of Standards and was subsequently published as an ASHRAE (1977) standard. In West Germany a mixed indoor/outdoor test method was produced by the Bundesverband Solarenergie (1978) and the Commission of the European Communities (1980) has studied both these methods in order to produce recommendations for European solar collector test methods. The British Standard Institution (1981) has published a draft method of test including both a steady state test for use in a solar simulator,

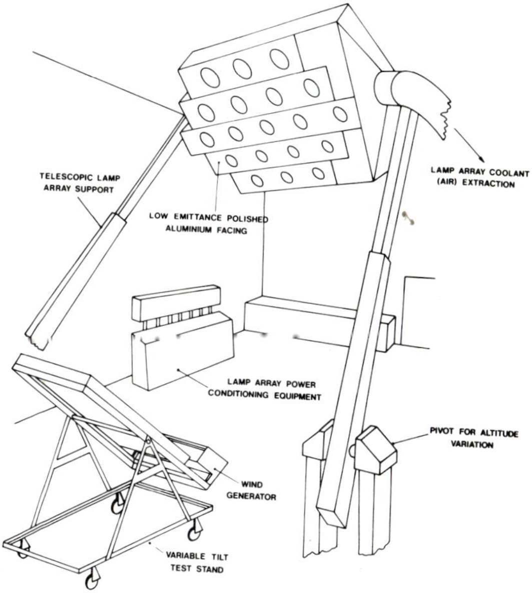

Figure 3.1.4: Solar simulator. This indoor test installation is used at the Solar Energy Unit, Cardiff.

as for example the one illustrated in Figure 3.1.4 which is used at the Solar Energy Unit, Cardiff, and a transient method for use under variable outdoor irradiance.

The durability of collectors has been a subject of a study by Zerlaut (1980 B) and by Hayward (1980) and Rawcliffe. Major surveys of existing solar heating institutions and the development of durability test methods are being carried out within the collaborative programmes of both the CEC and IEA.

The early development of the collector manufacturing industry in the UK was described by McVeigh (1979) and Batty (1980) is developing methodologies which can help to foretell the probable ways of market penetration by the new technology and its products. The task is made difficult by the fact that the traditional logistic growth curves do not work too well in a market which is considerably influenced by government intervention and economic uncertainty.

3.2 **Air Collectors**

Air heating collectors deserve a special mention. They offer the advantages over aqueous systems that they don't freeze and they don't create problems if leaks occur. However, their disadvantages are that if they do leak this may go undetected resulting in lower efficiency and they require large surface areas and high air velocities to achieve reasonable heat transfer. The first purpose built solar air collectors were designed by Dr. Maria Telkes for a house which was built in 1948 in Dover, Massachusetts. These vertical collectors were coupled with Glauber's salt phase change storage. A few years later, in 1957, Lof built a house in Denver, Colorado, with what has become one of the most famous solar air heating systems, mainly because it has worked so well for such a long period of time. Lof still lives in this house and monitors the well instrumented installation. Air heating collectors have gone through some changes in the mean time. A recent summary by Lof (1980) suggests design targets and many new designs have emerged from the USA, Japan, Sweden, Germany and the UK. A new matrix type solar collector was developed in Madison, Wisconsin, after an intensive five year period of research and analysis by Chiou (1966) and El-Wakil, Beckman (1968), and Hamid (1971) and Beckman into the details of the design of porous matrices and screen matrices.

A compact air-to-water heat exchanger based on heat pipes for use in connection with 6 m^2 flat plate solar air heating collectors was reported by Yaneske (1980) and Wiles, while an integrated collector and interseasonal heat storage system employing horizontal air heating collectors, mounted directly above a buried pebble bed store was described by Jones (1980) and Oreszczyn.

Air heating collectors used in grain drying is the subject of research and development work at the Agricultural Science Service, Slough Laboratory. Burrell (1979) has described the work done during the last two decades. Another agricultural application in Germany described by Boettcher (1980) and Delyannis uses polythene films to reduce frictional resistance, to improve insulation characteristics and to reduce the cost of the new air heating solar collectors. A stagnation temperature of 90°C in Germany and 120°C in India has been survived by these collectors without damage. The use of a static air pillow over the polythene film can increase the efficiency.

At Strathclyde University Yaneske (1981) and Wiles are investigating the optimisation of the relation between pressure loss and heat transfer efficiency in air heating collectors with flat broad mesh absorbers.

Although air heating collectors seem to have progressed slowly in their development it is noticeable that an interest has been raised partly as an adjunct to the renewed interest in passive solar heating.

3.3 Selective Coatings

Selective coatings can be used to reduce the transfer of heat by thermal radiation between the absorber and the cover in a solar collector and thus reduce the collection heat losses.

A coating applied to the cover must be capable of transmitting solar radiation whilst reflecting the thermal radiation emitted by the absorber. This is possible in principle because as discussed in Chapter 2 the spectrum of radiation emitted by a black body does not overlap significantly with that of solar radiation. The University of Waterloo lecture by Tabor (1979) gives a list of references to suitable materials. Care has to be taken that the transmittance of the coating is high enough not to impair the cover transmission.

A selective coating applied to an absorber must be capable of absrobing solar radiation whilst having a low emittance to thermal radiation. A typical spectral reflectance curve for a selective absorber surface is shown in Figure 3.3.1.

The first application of selective surfaces came from Tabor (1955) who used black nickel which became known all over the world as a "Tabor coating". These have been widely used in Israel for many years. An exhaustive survey of the optical properties of various surfaces is given by Meinel (1977) and Meinel while Seraphin (1979) and Zerlaut (1980A) give up to date surveys of fundamental material considerations for selective coatings. Duffie (1980) and Beckman devote considerable attention to radiation characteristics of opaque materials and give much practical information on radiation properties. The introductory paper by Tabor (1977) in the ASHRAE report gives a systematic guide through the theory and also mentions the interesting application of selective "cold" surfaces. These have high emissivity and low solar absorptivity for purposes of cooling and air-conditioning.

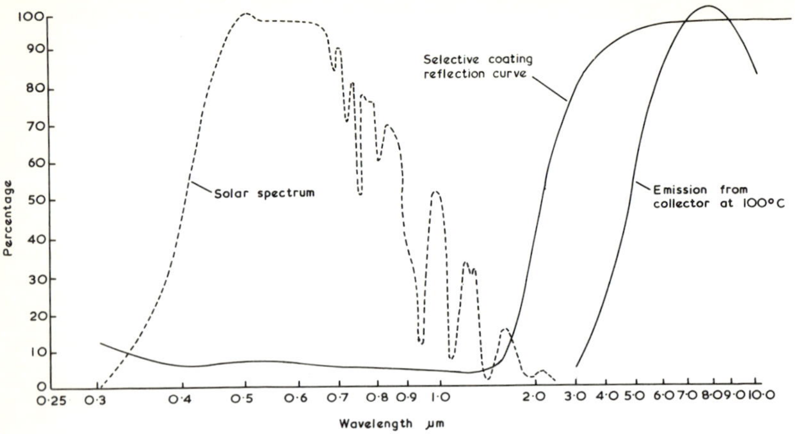

Figure 3.3.1: Selective absorber spectral distribution. Short wavelength radiation is transmitted, long wavelength radiation is reflected.

Selective surfaces can be produced by depositing a thin layer of black dielectric onto a polished metal substrate, for example black chromium onto polished nickel or black nickel onto iron. The physical structure of selective surfaces and the mechanism whereby they operate have been studied by Spitz (1979) and co-workers, and their optical properties by Behagel (1979) and co-workers. In simple terms this type of coating can be considered as a highly polished reflector for long wave thermal radiation while the black surface remains able to absorb the short wavelength solar radiation. The thickness of the black dielectric coating is critical to the selectivity and typically lies in the region of 1 micron. Such thin films can be produced by carefully controlled electro deposition or chemical dipping.

An alternative type of selective surface involves a sequence of layers deposited in such a way that they behave as a selective filter. Seraphin (1979) describes such surfaces which have been restricted to high temperature applications because of their high cost. In Japan, Ohno (1979) and co-workers at Nippon Steel reported an improvement of optical reflectance of a homogeneous interference film used for solar absorbers. They used texture control by selecting the principal components of the film such that optical matching with the substrate is ensured. This enables them to control the wavelength at which the reflectance spectrum starts rising.

Microscopic grooves or conical needles of a size similar to that of the wavelength of solar radiation can be deposited onto a surface to create selectivity. These are called wavefront discriminating materials and work on the principle that they reflect the wavelengths which are greater than the surface roughness. The size and shape of the surface is critical.

The cost of selective surfaces for collectors produced in large quantities remains a problem. Most commercial applications favour chemical bath processes. In the UK black chrome coatings have been commercially available for 4 years. The largest plant producing selective coatings in Europe is the result of a cooperative programme between Dornier System GmbH and Oberflachenveredelung Uhl GmbH in Germany and uses an anodising process in a sequence of 14 successive operations in huge tanks. This plant is described by Stemmer (1980) features preheating of the electrolyte and a high degree of control automation. In the United States selective copper oxide on copper substrate is the basis of the Ebanol blackening process.

As an alternative to electrodeposition of a coating directly onto large absorbers an adhesive film with a selective coating may be used. A commercial example, available from a British subsidiary of INCO, is that of a nickel foil with a black nickel coating.

The optical properties of selective surfaces as measured by Hutchins (1981) on a wide range of commercial samples has shown that the uniformity is often poor. This shows that the quantity control is one of the major remaining problems in the large scale production of selective coatings.

The need to produce commercially viable and constant quality surfaces has moved the industry to encourage investigations into the thermal degradation of black chrome coatings as reported by Lampert (1979) and into the consequences of reducing the chromium concentration in the plating bath in order to improve the thermal stability of the surface as reported by Pettit (1979) and Sowell. Both investigated Harshaw's Chromonyx coating in the 600 K temperature range and concluded that the surfaces did not degrade significantly under accelerated aging tests. According to McDonald (1975) the durability of black chrome coatings is good and exposure to humidity does not degrade the optical properties. Recently Smith (1981) and Ignatiev have shown that the addition of molybdenum to black chromium produces selective coatings which are stable in air up to 425°C.

Although there is yet a reluctant acceptance of the benefits that selective surfaces can bring, this is mostly due to less demonstrable effects on low temperature collector performance. Early teething problems with different surfaces and low volume production caused flat plate collectors to be not obviously cost effective when treated with selective surface coating. This is likely to change in the near future and it will change even more rapidly for high temperature concentrating collector applications. As the understanding of the component material structures is improving, the idea mentioned by Hutchins (1981) that in many real surfaces a combination of several processes contributes to the solar absorptance, can lead to multiple stacked tandems of semiconductor absorber/reflector layers with greatly enhanced and purposely designed selectivity, bringing obvious economic benefits.

3.4 **Evacuated collectors**

To attain a higher temperature than can be produced by the flat plate collectors described in section 3.1 it is necessary to reduce the convective heat loss significantly. This can be achieved by evacuating the space between the absorber and the cover. However, flat plate geometry does not lend itself easily to this process and tubular designs have therefore been developed. There are several commercial designs of evacuated collectors and they vary mainly in the design of the absorber.

Perhaps the simplest design is to place the evacuated tube over the top of the absorber plate as shown in Figure 3.4.1. This is analogous to a double glazed collector with partial vacuum but has the disadvantage that the back losses are governed by the level of insulation.

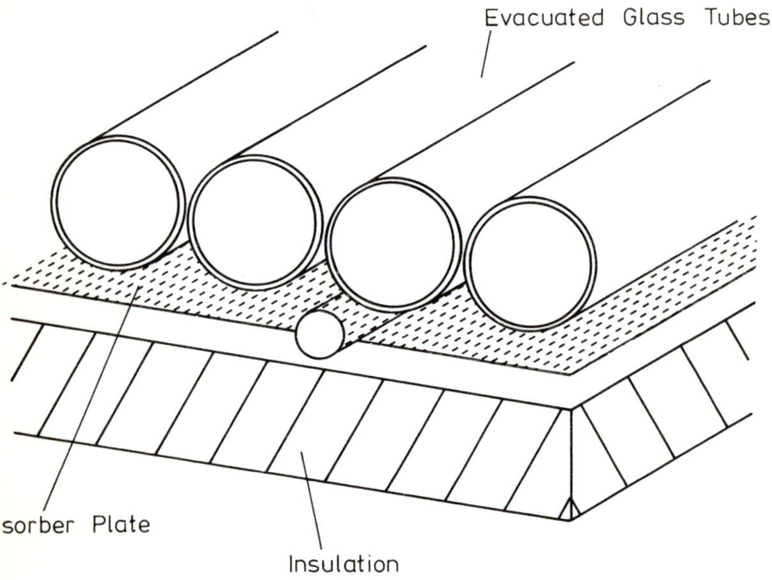

Figure 3.4.1: Evacuated tubes reduce convective heat loss from absorber plate, insulation reduces conductive heat loss.

To reduce heat losses further the absorber may be placed entirely in an evacuated environment. The absorber plate is cut into strips each one of which has a heat transfer tube bonded to it as shown in Figure 3.4.2. A disadvantage of this design is that it requires a glass to metal seal.

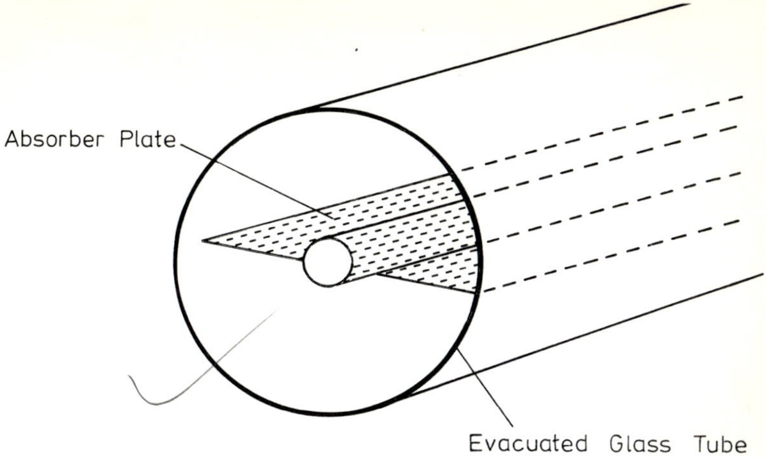

Figure 3.4.2: Absorber plate inside the evacuated glass tube. Glass to metal seals are needed at both ends.

To overcome the problem of glass to metal seals the design in Figure 3.4.3 may be used. Here a vacuum flask is placed over the absorber.

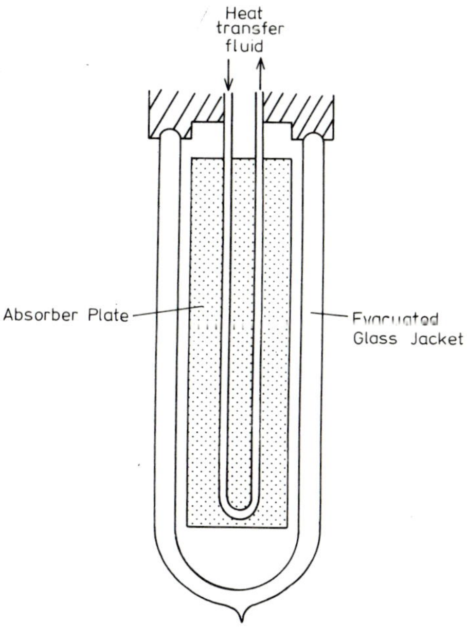

Figure 3.4.3: Transparent vcacuum flask reduces heat losses and requires seals at one end only.

Another means of avoiding the need for a glass-to-metal seal is to construct the absorber entirely out of glass using three concentric tubes. In the design shown in Figure 3.4.4 the heat transfer fluid enters through the central tube.

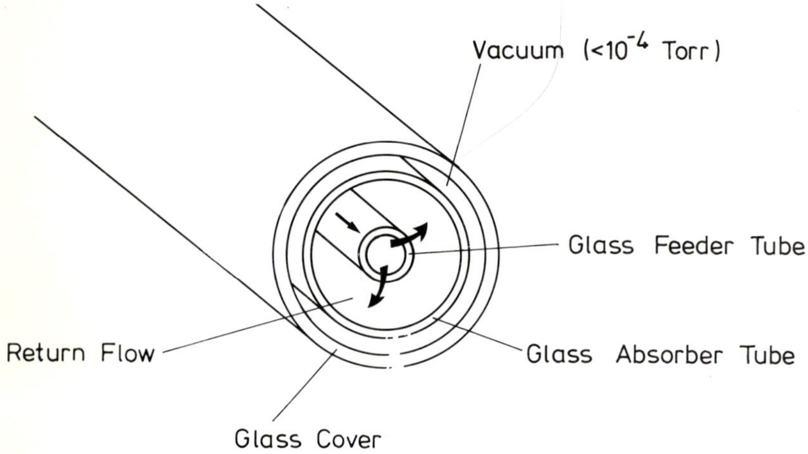

Figure 3.4.4: All glass evacuated tubular solar collector.

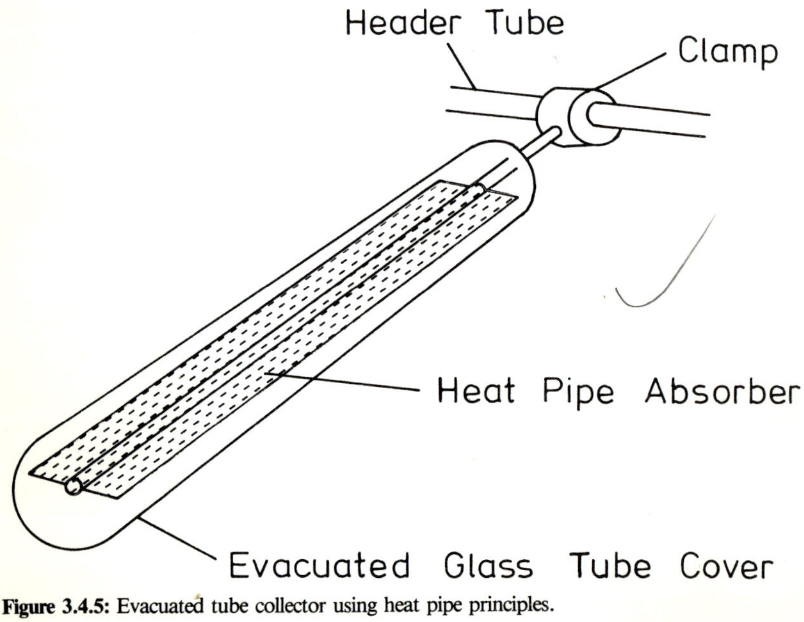

Figure 3.4.5: Evacuated tube collector using heat pipe principles.

A recent design described by Horster (1981) and Kersten and shown in Figure 3.4.5 combines a heat pipe with an evacuated tube collector. The selectively coated absorber is a fin which is attached to the centrally positioned heat pipe. A metal extension from this is clamped with a copper clamp to the hot water pipe. This guarantees that the primary heat transfer fluid is always sealed inside the heat pipe and heat is only transferred to the secondary circuit which carries the hot water. If a unit has to be replaced or repaired, it can be done without disturbing the rest of the system, thus whole banks of collector arrays can be built up in this modular way.

Evacuated collectors are commercially available in many parts of the world. Their proponents claim that in mass production they will become competitive with the best flat plate collectors while outperforming them at high temperatures. They are likely to be used increasingly in such applications as industrial process heating and air conditioning.

3.5 Concentrating collectors

The idea of concentrating solar radiation onto a small absorber area appealed to early inventors and today scientists are working on ways of producing elevated temperatures with solar energy using this technique. A French Professor Mouchot constructed a parabolic mirror which in the 1860's raised steam. Shuman in 1907 produced a design, which was basically a flat plate collector and had mirrors attached to it in order to achieve a concentration ratio of 2 : 1.

This however is not adequate for very high temperature applications for two reasons. Firstly, the concentration ratio, which can be seen as the ratio of the aperture area to the absorber area, limits the maximum operating temperature of the collector, and secondly, concentrating collectors can only accept a direct beam of solar radiation within a restricted acceptance angle (defined by the solid angle between the absorber and the aperture). Unless a concentrating collector can be moved to track the sun, variations in the incidence angle (between the direction of the solar beam and the normal to the plane of the collector aperture), will severely limit the time when the collector can direct the solar radiation onto its absorber.

A significant development in collector design sprang from the accidental discovery of the so-called 'compound parabolic concentrator' or CPC in 1966 by American and Soviet physicists working on high energy radiation research separately and simultaneously. These concentrators are non-imaging thus there is no need to move the collector to follow the sun precisely. The parabolic surfaces are arranged in such a way as to maximise the acceptance angle for a given concentration ratio. Their use for solar energy collection and a number of different geometries have been described by Rabl (1976, 1980). Welford (1978) and Winston have published a book which comprehensively describes the principles of non-imaging collectors and their applications.

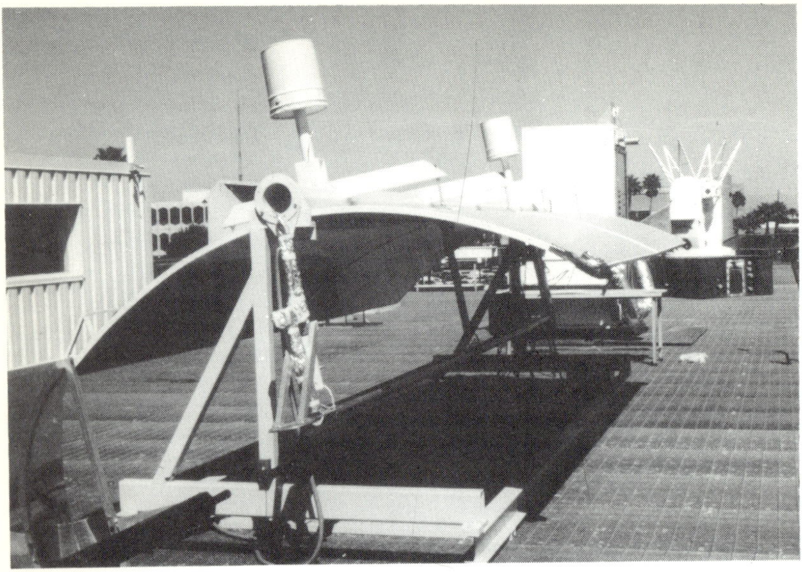

Figure 3.5.1: Light weight, one axis tracking, parabolic trough collector in parked position at Arizona State University.

One type of concentration utilizes a parabolic trough collector which comes in a variety of geometric arrangements. Some have a fixed focal line absorber and a moveable mirror. This makes the piping to the high temperature focal line easier and enables the mirror to be parked after sunset in an upside down position, which helps to avoid dust collecting on the mirror. A number of companies manufacture such collectors and the recent trends are towards light weight structures, for example of the honeycomb type, covered with a thin, tough, self-adhesive surface which can be peeled away if it is impaired through physical damage or optical deterioration. A typical example shown in Figure 3.5.1 uses an aluminium acrylic film bonded to 15 mm thick honeycomb structure and has a 0.9 m focal length. The trough is 3 m long and 1.4 m wide. The absorber is a selective black coated steel tube with a half cylinder Pyrex glass shield on the mirror side and reflective-insulating half cylinder shield on the top side, to reduce losses. The structure with mirror and absorber weighs little more than 50 kg/m^2, and is easy to transport and to assemble. Such collectors may be mounted east-west or north-south and tilted parallel to the polar axis. In either case the trough must be moved to follow the sun. When a very large trough is used it is possible to achieve the same results by moving the absorber pipe to follow the sun. The fixed trough must be aligned east-west and can be very long as shown in Figure 3.5.2.

Figure 3.5.2: Fixed trough, east-west collectors with movable absorber pipe at Saudia Laboratories in Albequerque, New Mexico.

A wide variety of concentrating collectors is described by Roy (1979) who discusses the different geometric configurations and gives practical details. The most recent and very comprehensive report by Rabl (1980) traces some historical developments and summarises the current theoretical knowledge on concentrating collectors. This authoritative account comes some 15 years after the discovery of CPC. Concentrating collectors were however discussed much earlier for example Laszlo described solar furnaces in the fifties, and Tabor (1958) calculated how much energy would be collected over the central part of the day by a large receiver whose trough and absorber are both fixed. Fixed concentrating collectors are now commercially available in modular form. Their collection efficiency is improved by covering the troughs with snap on plastic domes. Thus an entire roof surface can be covered and the collectors become an integral part of the roof structure.

Duffie (1962) and Lof proposed focusing solar collectors for power generation, and Lof (1963) and Duffie have worked on an optimisation method for a linear parabolic concentrator in which the reflecting mirror is divided into flat strips parallel to the focal line, similar to the ones shown in Figure 3.5.2 above.

In Egypt Sakr (1980) and El-Haggan developed an unusual concentrator configuration using a half paraboloid dish and a half parabolic cylinder. The heat absorber is shaped to suit the concentrator's energy distribution and heliostats are used to reflect radiation into a fixed stationary concentrator. The device generates steam with an overall efficiency of about 19%.

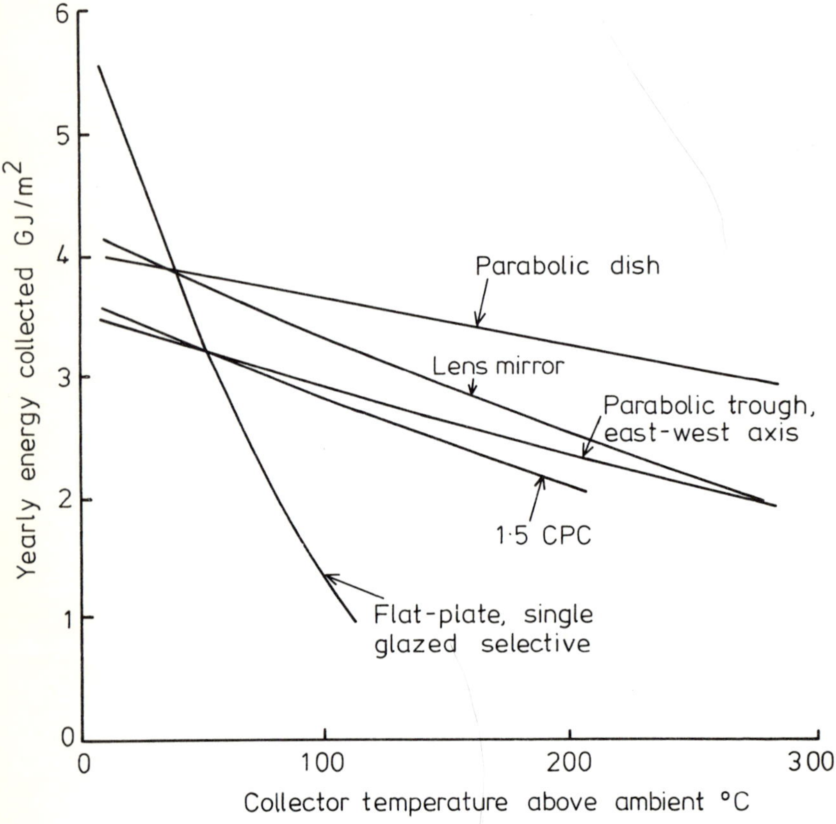

Figure 3.5.3: Annual energy collected versus collector temperature. Comparison of five types of collector, after Collares Pereira (1981).

The efficiency of flat plate and parabolic trough concentrating solar collectors were compared by Samuel (1980) and co-workers. Another comprehensive study was made recently by Collares Pereira (1981) who compared the yearly energy delivered by five different types of collector as a function of collector temperature above ambient. These are shown in Figure 3.5.3 and illustrate the point that at very low temperatures, say below 40°C, the flat plate collector has a higher efficiency. From above

50°C all types of concentrating collector gave higher efficiency than the flat plate collector. The figure also illustrates the fact that the slopes of their curves are shallower because concentrating collectors have lower heat losses.

3.6 **Gravitational solar thermal systems**

The most widely used and in many places the most obviously cost effective application of solar energy is in domestic water heating. The required installation can use components and technology from the central heating/plumbing trade, except for the collectors. By using flat plate collectors even this is brought within the competence of normal trade practices, although some special care has to be taken in fixing the collectors to the roof. Most mistakes in installations are basic ones and have nothing to do with solar technology. An excellent summary of what to look out for can be found in the Building Research Establishment Report by Wozniak (1979). Ensuring that air is removed from the system, that frost protection is provided, that proper water levels are maintained, that corrosion is avoided, that collectors are secure and roof mountings do not let rain through, that pipes and tanks are well insulated are the most important aspects of avoiding unnecessary malfunctions. A code of practice for designing, installing and operating systems has been published by the British Standards Institution (1980).

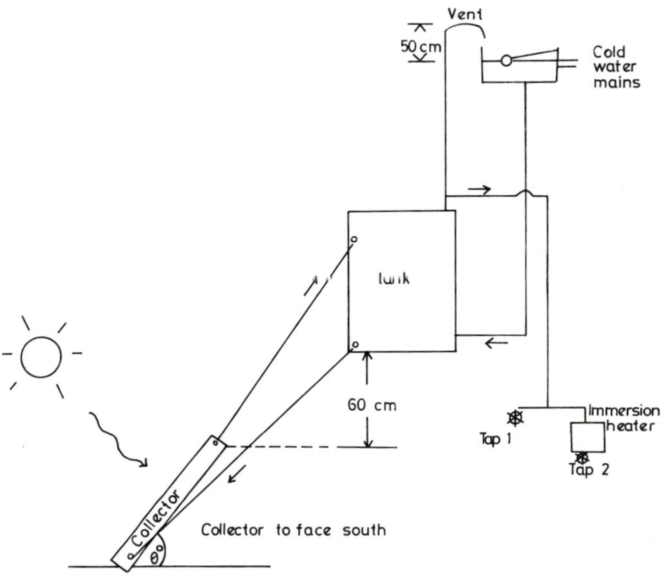

Figure 3.6.1: Gravitational solar water heating installation.

A gravitational solar water heater consists of a flat plate collector which is connected by a flow and return pipe to a tank placed above the collector. The return pipe is connected to the bottom of the tank and it brings the coldest water from the naturally stratified tank back to the collector. To make sure, the mains cold water feed is also connected near the bottom of the tank as shown on Figure 3.6.1. In this way the water flowing is always the coldest available and the collector operates with the highest efficiency. As the water in the collector is heated by solar radiation its density decreases and it will rise causing circulation. The flow pipe carrying the heated water from the collector is connected to the tank at a higher level than the return pipe. This ensures that the hot water is deposited in the upper part of the tank which is usually warmer because of natural stratification. Hot water can then be withdrawn from the upper part of the tank.

Figure 3.6.2: Close coupled tank and collector assembly on the flat roof of a private villa in Jerusalem.

This system is often referred to as a thermosyphon system as the circulation is driven only by the gravitational forces acting on the thermally inbalanced density distribution in the circuit. To guarantee a good circulation it is necessary to keep a head or height difference between the collector top and the tank bottom. Kreith (1978) and Kreider recommend a minimum of 0.3 m and Wozniak suggests 0.6 m to prevent reverse circulation although this can also be achieved by the use of non-return valves. The successful Israeli tank-and-collector assembly is shown in Figure 3.6.2. Experience shows that the diameter of the flow and return pipe as well as the construction of the collector has much to do with the resistance which the thermosyphon system has to overcome. The circulation will start after enough solar radiation has been received to raise the temperature of the water in the collector above a critical point which is different for every system. The thermal performance of the system depends on very delicate balances of energy received, water levels, frictional resistances, water temperatures, and the circulation is incapable of overcoming air locks, hence there is a need for large bore pipes and no sharp bends. The system cannot be drained easily and has no frost protection. For this reason it is more popular in the warmer climates such as Israel, Greece and Australia. Direct systems use the water which comes from the cold water feed line and therefore may be subject to corrosion.

To provide frost protection for gravitational systems it is possible to use an antifreeze solution in an indirect thermosyphon. This is different from the direct thermosyphon in two components, namely a heat exchanger and a header tank which are needed in addition to the components described above. Air locks can also occur in closed circuit systems and so they have to be vented as shown in Figure 3.6.3.

Indirect gravitational systems are more expensive and require more work and space during installation. They are subject to all the limitations listed for the direct systems and their main advantage of frost protection

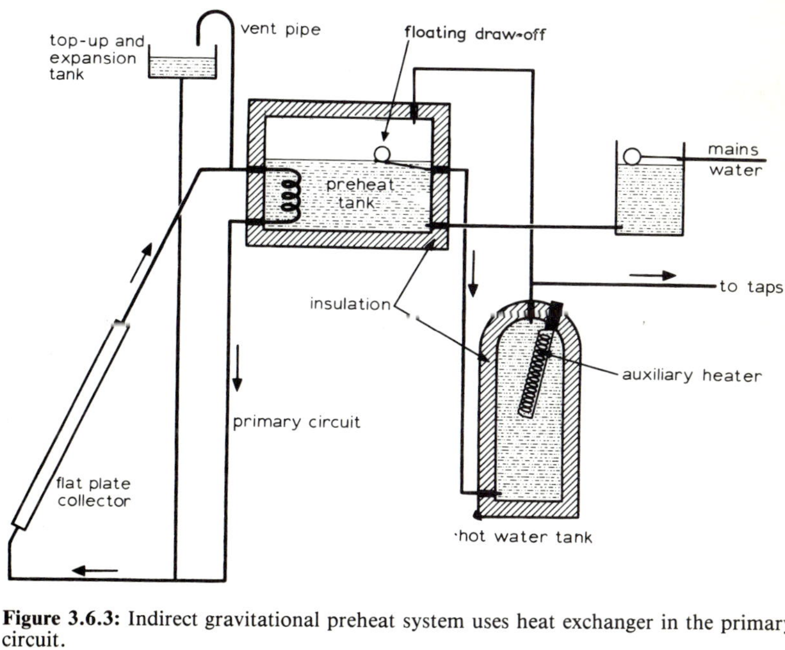

Figure 3.6.3: Indirect gravitational preheat system uses heat exchanger in the primary circuit.

is the primary justification for selecting them. Thermosyphon systems are popular because of their simplicity and low cost. They are particularly attractive in areas where there is no electricity supply such as in remote places, or where the supply of electricity is likely to be interrupted.

3.7 Pumped solar heating systems

To ensure circulation at the right time and at the right velocity pumps are often used. The ability to control the time when the circulation occurs is important to the overall system efficiency. By adding pumps to the system the velocity which will give high heat transfer rates can be guaranteed, however, pumps require the addition of a controller.

One of the simplest versions of the pumped system is the direct pumped circulation solar water heater with automatic drain-down for protection against freezing or overheating. The schematic diagram Figure 3.7.1 shows that the main components are the same as for thermosyphon system except for the pump, the controller and the different levels at which the

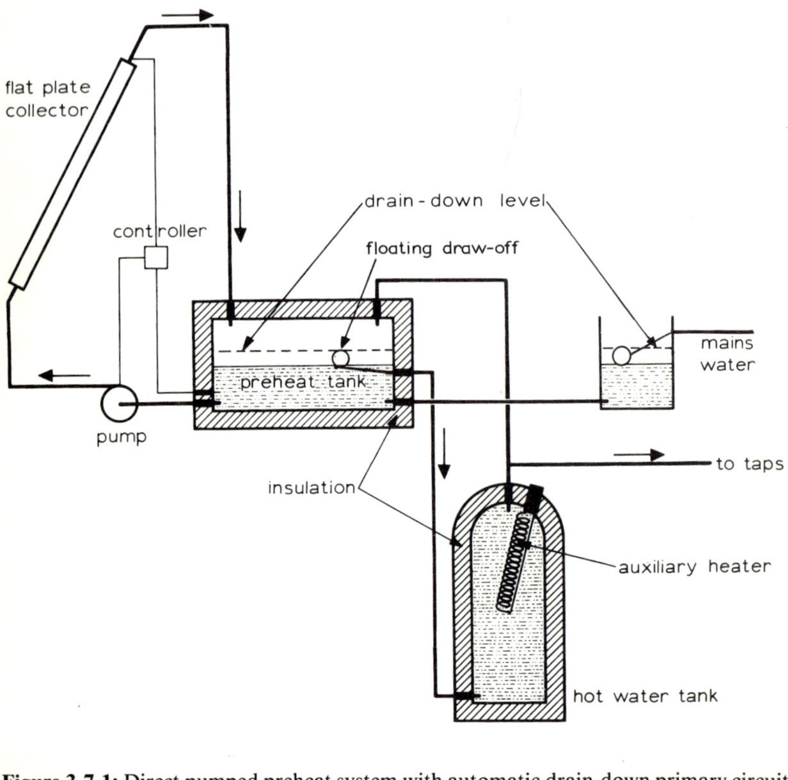

Figure 3.7.1: Direct pumped preheat system with automatic drain-down primary circuit.

components are located. The relative position of the tank and the collector now no longer matters, however, the pump needs to be at the lowest level so that it is always flooded. As long as the pump is running the collector will be full but when the pump stops water is drained down from the collector into the tank. A separate feed tank with a ball valve provides cold water from the mains supply at the same rate as the hot water is used up. An air vent at the highest point provides an exit for air when the pump starts up and the system is filled with water and it provides an inlet for air when the pump stops and the system is drained by gravity.

Another important component in these direct automatic drain-down systems is the controller. In most cases this senses the temperature difference between the outlet of the collector and the bottom of the storage tank and if this exceeds a certain value the pump is turned on, if the value is less than a preset figure the pump is turned off. The turn-on temperature difference is usually 3-4°C and the turn-off temperature difference is about 1°C. Malfunctions of systems can often be attributed to poor installation of sensors. The collector sensor put in the air space between the plate and the cover instead of being bonded to the absorber waterways will not register the temperature of the water in the channels or pipes. The performance of a sensor strapped onto the outside of the cylinder instead of being bonded to it will depend on how good the contact is between it and the curved cylinder wall. The result of using poorly mounted sensors could be that the pump runs for longer than is desirable.

Indirect pumped systems offer more flexibility to the designer. They have two circuits, a primary and a secondary. The primary circuit has in it the pump, the collector, the heat exchanger (usually inside the tank) and a make up tank which provides expansion and venting. The secondary circuit starts with the cold water supply, includes the solar tank and finishes in the hot water supply line. The auxiliary heat input which guarantees that the final temperature of the delivered hot water meets the design specification can take the form of an electric immersion heater installed in the solar tank. Usually the solar water heating system is only a water preheat system added on to an existing system of hot water supply. If so, there are two tanks, one is the solar preheat tank and the other is the existing domestic hot water tank. In this case a slight modification of the existing system is needed and the cold water feed has to come into the solar preheat tank's cold inlet connection. This arrangement shown in Figure 3.7.2 is the favoured design for retrofit domestic hot water solar preheat installations. The separation of the primary and secondary circuits allows the use of antifreeze or non-aqueous liquids in the collector loop. It also allows the use of materials which would otherwise be corroded by contact with fresh water.

For larger installations in buildings such as offices, hospitals, factory cafeterias, schools and old people's homes where the hot water demand is

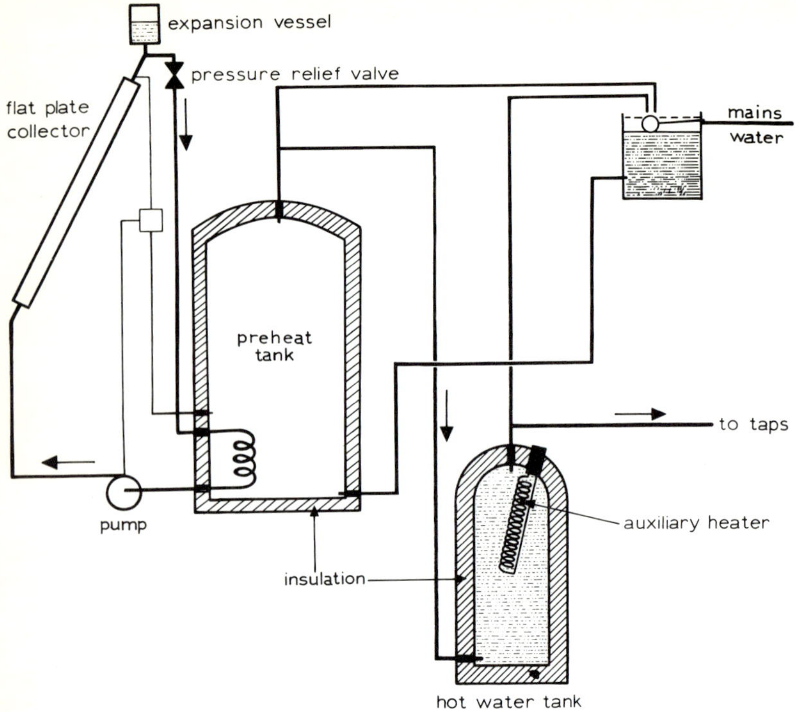

Figure 3.7.2: Indirect pumped preheat system with heat exchanger and controller.

larger but not of a specialised industrial nature it may pay to go for more elaborate systems. These could include pumps both in the primary and in the secondary circuits with an external heat exchanger connecting the two. More than one tank in the secondary circuit could also be employed. Variable flow control of the primary circuit brings a small but worthwhile advantage in system performance. This aspect has been examined by Brinkworth (1975) and (1980) who found it to be a low sensitivity factor.

An installation at an old people's home in Birmingham, UK, designed by Jesch includes the features of variable flow rates and distributed energy storage. The split system has half of its collectors facing SE and the other half facing SW but both separate and independent systems feed one common set of energy storage tanks. The storage tanks operate at different temperatures to increase the ability of the system to collect more solar energy and for longer periods of time. Jesch (1979, 1981 B, 1981 C) and co-workers have reported on various aspects of this scheme.

In Yugoslavia on the Adriatic coast in the resort area of Pula a 300 bed hotel is being built with solar water heating and swimming pool heating.

The system provides all energy requirements during the tourist season. On completion the total collector area will be 660 m^2 with two storage tanks, several heat exchangers, pumps, motorised three way valves, etc. The controls are designed to accommodate the possibility of adding a heat pump to the system, Vosilla (1980).

For space heating in northern latitudes the concept of an energy roof is gaining increased attention. In this type of system the whole roof of a building acts as a collector for solar energy and is often integrated with heat pumps and heat recovery system. Heidtmann (1980) gives interesting examples from Germany and Neal (1979) has reported on an experimental house in the U.K.

Experience with water heaters in the U.K. and other parts of Northern Europe show that significant energy can be saved even in these high latitudes. Demonstration projects are being perused both with national and CEC funds which will provide valuable data for future designs.

3.8 **Simulation and monitoring of pumped systems**

The first incentive for any design activity which uses computerised mathematical models is to optimise the whole system in operation. This requires the knowledge of a number of input parameters such as the environmental data, the likely irradiance at the location of the installation, the seasonal variations in air temperature, wind velocity, cold water temperature, etc. Most importantly the designer wants to know the user behaviour pattern. Economic considerations also come into play for they will decide whether to go for one collector with high efficiency but high capital cost or another with somewhat reduced efficiency but with lower capital cost.

The complexity of the analysis requires computers which can match the size and type of programs but both simple programs which can be executed on hand held calculators and comprehensive large models which need sophisticated computers have an important role to play.

The best known proponents of simulation and computer optimisation are Beckman (1977), Klein and Duffie. After the development of TRNSYS (1973) which requires large computers, they proposed a simplified method, the f-chart method, which became widely accepted in the solar community. This method correlates long term system performance with averaged climatic and other data. The published correlations were developed for American conditions but the method is universally applicable. It enables the designer to use a programmable pocket calculator to make very rapid estimates of system performance which are sufficiently accurate for most purposes. More recently (1980) they published a book which gives detailed instruction about preparing these programs. The idea of the f-charts was the subject of Beckman's (1979) keynote paper at the Birmingham UK-ISES meeting, where he also mentioned the utilizability of flat plate collectors on which Klein (1978) reported earlier.

At the Cardiff Solar Energy Unit, Brinkworth (1977) worked on auto-correlation and stochastic modelling of insolation sequences and published a paper on asymptotic behaviour as a guide to the long term performance of solar water heating systems. In his later paper (1978) he reported on his model in detail.

At the Solar Energy Laboratory in Birmingham University Jesch (1977, 1978) and Soldatos published their simulation model which has been used in a number of design cases. This program can deal with specialised storage and control aspects of complex active solar systems. Jesch (1981B) and Greeves showed the benefits to be expected from forced stratification in distributed storage tanks.

A Brunel University building's solar installation was designed by Khor (1979) and Smith using the f-chart method and Munroe (1980) reported an electric circuit simulation of the flat plate solar water heating system. Howarth (1980) uses a nomogram for estimating the performance of solar hot water heaters.

A comprehensive model which can deal with a large variety of active solar heating and cooling systems has been developed by Rosenfeld (1979) and reported on its use for simulating domestic water heaters performance in the U.K. Pye (1980) and Rosenfeld showed their simulations to be in good agreement with experimental results obtained from a field trial in London monitored by Makkar (1980). Pye (1981) and da Rocha have used the program to design a large solar water heater for an office canteen installation in Lisbon.

Michelson (1980 A) reported on the optimisation of a solar water heating system using the simplex method. Her simulations (1980 B) of the performance of a water heater serving a block of students flats agreed well with the results obtained by Taylor (1980) from the monitoring of the 25 m^2 installation.

Of the numerous other models available elsewhere the efforts of the Philips Research Laboratories at Aachen in Germany should be mentioned. There Horster and his team are working on a number of major modelling developments, Horster (1980).

At Ann Arbor, Michigan, a group has developed several models. Clark (1980) reports on a life-cycle model which determines the annual performance in a single calculation step. Another of their models is described by Nabozny (1980) and co-workers. This is an interactive hourly model for air-systems with step-by-step instructions for the user.

The validation of existing models is an important reason for monitoring the performance of real systems. The monitoring must be at the level of complexity required to test the program. This often entails the use of frequent logging of data obtained from sensors under computer control, including climatic variables, temperatures and flow rates in various parts of the system. For convenience these data may be transmitted back to the laboratory by a dedicated landline for analysis.

Larger systems not only can afford but also need more sophisticated controls. These controls are likely to double as monitoring devices as well since the information from monitoring is very valuable for future design and setting of the controls themselves. Solar radiation, wind velocity, air temperature, fluid temperatures in many parts of primary and secondary circuits, cold water feed temperatures, cold water flow rate, hot water temperatures, tank temperatures, pump operating times are the most important control/monitoring parameters. All these data have to be monitored and controlled as a function of time. It is necessary to know when the pump was turned on or off and what was the radiation, flow rate, etc., at that time.

In complex systems both controlling and monitoring could be performed by one computer which, depending on the level of its intelligence, can also do the evaluation and readjustment of control parameters. Such heuristic systems have been reported by Jesch (1979) and Zekios who outlined the concepts and gave some monitored data, such as the diurnal and seasonal water usage variations, in a multioccupant home for the elderly. More recently Jesch (1981 C) and co-workers reported on control strategies which affect the overall operating efficiency of the system.

Tanaka (1980) and Komiya have reported on a project to determine cost effective, accurate and reliable monitoring methods. A highly instrumented experimental house near Osaka, Japan, incorporating active solar heating, air conditioning and water heating systems, has been used not only to compare its performance to simulation predictions but also to compare different types of sensors and instrumentation.

The development of simulation programs and the monitoring of systems in the field is being actively pursued in international programmes coordinated by the CEC and the IEA and by many national programmes including the one funded through ETSU in the UK. It is hoped that the modelling and simulation efforts will lead to a better understanding of how active solar systems work and how to design them to work better and cheaper.

References

ASHRAE, (1977): Method of testing to determine the thermal performance of solar collectors (ASHRAE 93-77), 1977.

Batty, S. S. (1980: "Methods of estimating solar market penetration" in **Solar Energy in the 80's.** Conference (M2) Birmingham, UK-ISES Midlands Branch 1980, 98-103.

Beckman, W. A. (1968): "Radiation and convection heat transfer in a porous bed". **Trans. ASME Journal of Engineering for Power 90** 1968, 51.

Beckman, W. A. (1977), S. A. Klein and J. A. Duffie: **"Solar Heating Design by the f-chart method".** Wiley, New York, 1977.

Beckman, W. A. (1979): "Design procedure for industrial process heating and cooling with solar energy". in **Practical Application of Solar Energy in Industry and Commerce.** Conference (M1), Birmingham, UK-ISES Midlands Branch, 1979, 24-30.

Behaghel, J. M. (1979), S. Berthies, J. Lafait and J. Rizzi: "Chromium black coatings for photochemical conversion of solar energy part II: optical properties". **Solar Energy Materials 1,** 1979, 201-213.

Boettcher, A. (1980) and E. Delyannis: "Entwicklung eines solaren Luftkollektors" in **Statusbericht Sonnenenergie** I, VDI-Verlag 1980, 95-101.

Brinkworth, B. J. (1975): "Selection of design parameters for closed circuit forced circulation solar heating systems". **Solar Energy 17,** 1975, 331-333.

Brinkworth, B. J. (1977): "Autocorrelation and stochastic modelling of insolation sequences". **Solar Energy 19,** 1977, 343-347.

Brinkworth, B. J. (1978): "Asymptotic behaviour as a guide to the long term performance of solar water heating systems". **Solar Energy 21,** 1978, 171-175.

Brinkworth, B. J. (1980): "The British Standard Code of Practice and its implications for solar hot water system design". in **Solar Energy in the 80's.** Conference (M2) Birmingham, UK-ISES Midlands Branch, 1980, 79-84.

British Standard Institution Code of Practice (1980): **Solar Heating Systems for Domestic Hot Water.** BS 5918:1980, 17 p.

British Standard Institution, (1981): **Methods of tests for the thermal performance of solar collectors".** No. IS DD 77 BSI, London 1981.

Bundesverband Solarenergie (1978): **Usability of Solar Collectors,** BSE, 1978.

Burrell, N. J. (1979): Grain drying using solar energy "in **Practical Applications of Solar Energy in Industry and Commerce.** Conference (M1), Birmingham, UK-ISES Midlands Branch 1979, 39-47.

CEC (1980): Recommendations for European Solar Collector Testing Methods, edited by A. Derrick and W. B. Gillett, Commission of the European Communities DG XII, 1980.

Chiou, J. P. (1966) and M. M. El-Wakil: "Heat transfer and flow characteristics of porous matrices with radiation as a heat source". **Trans. ASME Journal of Heat Transfer 88,** 1966, 69.

Clark, J. A. (1980): "A life-cycle model for solar heating system design and economic evaluation – MESH". in Proc. of the Solar Energy and Conservation Symposium, 11-13 Dec. 1978, Miami Beach, Florida. **Solar Energy and Conservation.** ed. T. Nejat Veziroglu. Pergamon Press, 1980. Vol. 3, 1347-1358.

Collares Pereira, M. (1981): "CPC-type concentrators and their applications, comparison with other collector types". in **Recent Developments in Solar Collector Design.** Conference (C25) London, UK-ISES 1981, 76-84.

Duffie, J. A. (1962) and G. O. G. Lof: Focusing solar collectors for power generation. **World Power Conference,** Melbourne 1962. Paper 207, III 7/5.

Duffie, J. A. (1980) and W. A. Beckman: **Solar Engineering of Thermal Processes.** Wiley-Interscience 1980, 762 p.

Gillett, W. B. (1981) and M. G. Hutchins: "Heating Water by the Sun". UK-ISES 1981.

Hamid, Y. H. (1971) and W. A. Beckman: "Performance of air-cooled radiatively heated screen matrices". **Trans. ASME, Journal of Engineering for Power.**

Hayward, K. (1980) and R. W. Rawcliffe: "Durability testing of solar collectors". in **Solar Energy in the 80's.** Conference (M2) Birmingham, UK-ISES Midlands Branch 1980, 50-55.

Heidtmann, U. (1980): "Aufbau und Funktion von Energiedächern":in **Tagungsbericht des 3. Internationalen Sonnenforums—Freie Vorträge,** Hamburg 1980. GDS—Sonnenenergie Verlag, 178-187.

Heywood, H. (1954): "Solar energy for water and space heating". **Journal Inst. of Fuel 27,** 1954, 334-352.

Hollands, K. G. T. (1965): "Honeycomb devices in flat-plate solar collectors". **Solar Energy 9,** 1965, 159.

Hollands, K. G. T. (1976), T. E. Unny, G. D. Raithby and L. Konicek: "Free convection heat transfer across inclined air layers". **Trans. ASME: Journal of Heat Transfer 98,** 1976, 189-193.

Hollands, K. G. T. (1978), K. N. Marshall and R. K. Wedel: "An approximate equation for predicting the solar transmittance of transparent honeycombs". **Solar Energy 21,** 1978, 231.

Hollands, K. G. T. (1979): "Free convection in solar collectors". in **Solar Energy Conversion.** (5th Course, Univ. of Waterloo Canada 1978) ed. A. E. Dixon and J. D. Leslie, Pergamon Press 1979, 126-147.

Hottel, H. C. (1942) and B. B. Woertz: "Performance of flat-plate solar-heat collectors". **Trans. American Society Mechanical Engineers 64.**

Hottel, H. C. (1959) and T. A. Unger: "The properties of a copper oxide-aluminium selective black surface absorber of solar energy". **Solar Energy 3** (3) 1959, 10.

Horster, H. (1980): **Wege zum energiesparenden Wohnhaus.** Philips Fachbucher, Hamburg, 1980.

Horster, H. (1981) and R. Kersten: "Evacuated solar collectors". in **Recent Developments in Solar Collector Design** Conference (C25) London.

Howarth, A. T. (1980): "A nomogram for estimating the performance of hot water heaters". in **Modelling and Monitoring Solar Thermal Systems** in Conference.

Hutchins, M. G. (1981): "Recent advances in selective surfaces". in **Recent Developments in Solar Collector Design.** Conference (C25) London, UK-ISES 1981, 43-50.

Jesch, L. F. (1977) and P. Soldatos: "Solar water heating, engineering and economic analysis of a new technology". University of Birmingham, Dept. of Industrial Economics and Business Studies, Series B, No. 47. 1977, 29 pp.

Jesch, L. F. (1978) and P. G. Soldatos: "Solar water heating performance analysis by mathematical modelling". Research Report No. 166. University of Birmingham, Mechanical Engineering Department, 1978, 15 pp.

Jesch, L. F. (1979) and A. Zekios: "Adaptive control for a multiple tank, multiple temperature solar energy storage system". Proc. ISES Silver Jubilee Congress in Atlanta, 1979, Vol. 1, 852-856.

Jesch, L. F. (1981 A), U. Elani and T. E. Naybours: "Diagnostic tools for flat plate solar collectors". in **Recent Developments in Solar Collector Design.** Conference (C25) London, UK-ISES 1981, 1-22.

Jesch, L. F. (1981 B) and T. W. Greeves: "Solar energy storage in forced stratification tanks". in **Energy Storage.** Papers presented at the International Conference in Birmingham, UK, April 1981, BHRA Fluid Engineering, 1981, Vol. 1, 247-257.

Jesch, L. F. (1981 C), T. W. Greeves and D. Haralambopoulos. "Simulation of optimal control strategies for large solar water preheat systems". Proc. ISES Solar World Forum, Brighton, UK, 1981.

Jones, B. W. (1980) and T. Oreszczyn: "An integrated collector/interseasonal store of solar energy (Prometheus)". in **Solar Energy in the 80's.** Conference (M2) Birmingham, UK-ISES, 1980, 38-43.

Kenna, J. P. (1981): "Materials and design methods to improve the performance of flat plate collectors". in **Recent Developments in Solar Collector Design.** Conference (C25) London, UK-ISES 1981, 23-42.

Khor, C. L. (1979) and B. E. Smith: "Outline design of a solar assisted hot water system for a University physics building using the f-chart method". in **Practical Applications of Solar Energy in Industry and Commerce.** Conference (M1) Birmingham, UK-ISES Midlands Branch, 1979, 87-90.

Klein, S. A. (1974), J. A. Duffie and W. A. Beckman: "Transient considerations of flat plate solar collectors". **Trans. ASME Journal of Engineering for Power 96** 1974, 109-113.

Klein, S. A. (1978): "Calculation of flat plate collector utilizability". **Solar Energy 21** (5), 393, 1978.

Klein, S. A. (1979) and W. A. Beckman: "A general design method for closed loop solar energy systems". **Solar Energy 22,** (3) 269, 1979.

Klein, S. A. (1980) and W. A. Beckman: Solar Energy Programs for HP-41 and TI-59 Calculators. Solar Energy Laboratory, University of Wisconsin, Madison, 1980.

Kreith, F. (1978) and J. E. Kreider: **Principles of Solar Engineering.** McGraw-Hill, 1978, 778 p.

Lampert, C. M. (1979): "Thermal degradation of a black chrome solar selective absorber coating: short term". Proc. ISES Silver Jubilee Congress in Atlanta 1979, Vol. 3, 1892-1896.

Lof, G. O. G. (1963) and J. A. Duffie: "Optimization of focusing solar-collector design". **Journal of Engineering for Power.** 85A 1963, 221.

Lof, G. O. G. (1980): "Flat-plate and nonconcentrating collectors". in **Solar Energy Technology Handbook Part A Engineering Fundamentals.** ed. W. C. Dickinson and P. N. Cheremisinoff, Marcel Dekker 1980, 217-255.

Makkar, L. (1980): "Monitoring the three tap and hybrid solar water heating systems at Whateley Road, London". in **Modelling and Monitoring Solar Thermal Systems.** (C24) London, UK-ISES, October 1980, 30-39.

Mar, H. Y. B. (1975), J. H. Lin, P. P. Zimmer, R. E. Peterson and J. S. Cross: "Optical coatings for flat plate solar collectors". ERDA Report, Contract No. NSF-C-957, Sept. 1975.

McDonald, G. E. (1975): "Spectral reflectance properties of black chrome for use as a solar selective coating". **Solar Energy 17,** 1975, 119.

McVeigh, J. C. (1979): "Success or failure? Some factors influencing the development of the solar water heating industry in the United Kingdom". in **Practical Applications of Solar Energy in Industry and Commerce.** Conference (M1) Birmingham, UK-ISES, Midlands Branch, 1979, 104-108.

Meinel, A. B. (1977) and M. P. Meinel: **Applied Solar Energy,** Addison-Wesley 1977, 651 p.

Michelson, E. (1980 A): "Optimisation of a solar water heating system using the Simplex method". Internal report, Mech. Eng. Dept., University College, Cardiff.

Michelson, E. (1980 B): "A simulation of the solar water heating system at Llys Tal-y-Bont'. Internal report, Mech. Eng. Dept., University College, Cardiff.

Munroe, M. M. (1980): "Electric circuit simulation of the flat plate solar water heating system". in **Solar Energy in the 80's, Technical and economic viability.** Conference (M2) Birmingham, UK-ISES Midlands Branch, 1980, 131-137.

Nabozny, R. L. (1980), J. A. Clark and M. A. Egenka: "An hour model for design and economic optimization of solar space and water heating systems. Air system-SOLHOUSE II". in Proc. of the Solar Energy and Conservation Symposium, 11-13 Dec. 1978, Miami Beach, Florida. **Solar Energy and Conservation.** ed. T. Nejat Veziroglu.

Neal, W. E. J. (1979), D. L. Loveday and M. Pabon-Diaz: "A solar assisted heat pump and storage system for domestic space and water heating using a conventional roof as a radiation absorber". Proc. ISES Silver Jubilee Congress in Atlanta, 1979, **1,** 822-826.

Ohno, J. (1979), Y. Shindoh, J. Oka and H. Okada: "Improvement of optical reflectance of homogeneous interference film for solar absorber by texture control". Proc. ISES Silver Jubilee Congress in Atlanta, 1979. Vol. 3, 1902-1906.

Pettit, R. B. (1979) and R. R. Sowell: "Thermal ageing characteristics of electrodeposited black chrome solar coatings". Proc. ISES Silver Jubilee Congress in Atlanta 1979, Vol. 3, 1897-1901.

Pye, D. B. (1981) and A. Vianna da Rocha: The design of a solar water heating system for a Lisbon Canteen. Proc. Solar World Forum Congress, ISES, Brighton, 1981.

Pye, D. B. (1980) and J. L. J. Rosenfeld: "The mathematical modelling of active solar water and space heating systems". in **Modelling and monitoring solar thermal systems.** Conference C24 at the Geological Society, London, UK-ISES October 1980, 12-29.

Rabl, A. (1976): "Comparison of solar concentrators". **Solar Energy 18,** 1976, 93.

Rabl, A. (1980): "Concentrating collectors". in **Solar Energy Technology Handbook. Part A Engineering Fundamentals.** ed. W. C. Dickinson and P. N. Cheremisinoff, Marcel Dekker 1980, 257-343.

Rogers, B. A. (1979): **Data Analysis for Transient Collector Testing.** Univ. College, Cardiff, Dept. of Mech. Eng. and Energy Studies Report No. 517, June 1979. SEU No. 130.

Rosenfeld, J. L. J. (1979): "The thermal performance of solar water heaters in the UK". Proc. ISES Silver Jubilee Congress in Atlanta 1979, Vol. 2 977-982.

Roy, A. S. (1979): "Concentrating collectors". in **Solar Energy Conversion** (5th Course, Univ. of Waterloo, Canada, 1978) ed. A. E. Dixon and J. D. Leslie, Pergamon Press 1979, 185-252.

Sakr, I. A. (1980) and A. M. El-Haggan: "Peacock tail concentrator system for steam generation". in Proc. of the Solar Energy Conservation Symposium, 11-13 Dec. 1978, Miami Beach, Florida. **Solar Energy and Conservation,** ed. T. Nejat Veziroglu. Pergamon Press 1980, Vol. 1, 141-153.

Samuel, A. A. (1980), U. S. P. Shet, V. Sriramulu and M. C. Gupta: "Flat plate collectors and east-west oriented parabolic trough concentrators – option for solar thermal power plants". in Proc. of the Solar Energy and Conservation Symposium, 11-13 Dec. 1978, Miami Beach, Florida. **Solar Energy and Conservation,** ed. T. Nejat Veziroglu. Pergamon Press 1980, Vol. 1 33-40.

Seraphin, B. O. (1979): "Spectrally selective surfaces in photothermal solar energy conversion". in **Solar Energy Conversion.** (5th Course, Univ. of Waterloo, Canada 1978) ed. A. E. Dixon and J. D. Leslie, Pergamon Press 1979, 287-329.

Shewen, E. C. (1979) and K. G. T. Hollands: "Equations for representing the UL dependence in collector test procedures". Proc. ISES Silver Jubilee Congress in Atlanta, 1979, Vol. 1, 360-364.

Smith, G. B. (1981) and A. Ignatiev: "Black chromium – molybdenum: a new stable solar absorber". **Solar Energy Materials, 4,** 1981, 119-133.

Spitz, J. (1979), T. Van Danh and A. Aubert: "Chromium black coatings for photothermal conversion of solar energy, Part 1: preparation and structural characterisation". **Solar Energy Materials, 1,** 1979, 189-200.

Stemmer, R. (1980): "Entwicklung einer Fertigungsanlage zur selektiven Beschichtung von Aluminium-Solarabsorbern". in **Status bericht Sonnenenergie 1,** VDI-Verlag 1980, 227-233.

Tabor, H. (1955): "Solar energy collector design with special reference to selective radiation". **Bull Res. Counc. of Israel 5C,** 5-27.

Tabor, H. (1958): "Stationary mirror systems for solar collectors". **Solar Energy 2,** 1958, 27-36.

Tabor, H. (1977): "Selective surfaces for solar collectors". in **Applications of Solar Energy for Heating and Cooling of Buildings.** ed. R. C. Jordan and Y. H. Liu. ASHRAE GRP 170, VI 1-10, 1977.

Tabor, H. (1979): "Selective surfaces". in **Solar Energy Conversion** (5th Course, Univ. of Waterloo, Canada 1978) ed. A. E. Dixon and J. D. Leslie, Pergamon Press 1979, 253-286.

Tanaka, I. (1980) and H. Komiya: "Experimentierhaus fur die Nutzung der Sonnenenergie in Japan". in **Tagungsbericht des 3. Internationalen Sonnenforums,** 24 June, 1980, Hamburg, 198-207.

Taylor, P. J. (1980): "Monitoring a large solar water-heating system". **Solar Energy in the 80's** Conference (M2) Birmingham, UK-ISES Midlands Branch, 1980, 62-67.

TRNSYS (1973): "A transient simulation program, user's manual". Engineering Experiment Station. Report No. 38, Solar Energy Laboratory, University of Wisconsin-Madison.

Vosilla, S. (1980): "Utilisation of solar energy in the hotel of ZB-Banjole, Pula, Yugoslavia". in **Tagungsbericht des 3, Internationalen Sonnenforums,** 24 June, 1980, Hamburg, 208-222.

Welford, W. T. (1978) and R. Winston: "The optics of non-imaging concentrators, light and solar energy". Academic Press, 1978, 200 p.

Wozniak, S. J. (1979): **Solar Heating Systems for the UK: Design, Installation and Economic Aspects.** Building Research Establishment, HMSO, 1979, 100p.

Yaneske, P. P. (1980) and I. C. Wiles: ''The development of an air-to-water heat exchanger for the supply of domestic hot water from air-heating solar collectors''. in **Solar Energy in the 80's.** Conference (M2) Birmingham, UK-ISES Midlands Branch 1980, 56-61.

Yaneske, P. P. (1981) and I. C. Wiles: ''Air collectors''. in **Recent Developments in Solar Collector Design.** Conference (C25) London, UK-ISES 1981, 63-75.

Zerlaut, G. A. (1980 A): ''Fundamental materials considerations for solar collectors''. in **Solar Energy Technology Handbook. Part A, Engineering Fundamentals.** ed. W. C. Dickinson and P. N. Cheremisinoff, Marcel Dekker, 1980, 403-437.

Zerlaut, G. A. (1980 B): ''Outdoor testing of solar materials and solar collectors''. in **Solar Energy Technology Handbook. Part A, Engineering Fundamentals.** ed. W. C. Dickinson and P. N. Chemerisinoff, Marcel Dekker 1980, 439-456.

Chapter 4

THERMAL STORAGE OF SOLAR ENERGY

Whereas the need for air-conditioning coincides with an abundance of solar radiation, the demand for energy is often out of phase with the availability of solar radiation as for example space heating in winter. Generally the availability of solar energy and the need for this energy do not overlap sufficiently to allow systems to operate without some form of storage. The storage of solar energy collected by passive means was mentioned in Chapter 2. The present chapter covers the use of sensible and latent heat storage for use with active collection systems.

Since the purpose of thermal storage is to balance the supply and demand of energy, the size of the thermal store must be based on the estimates of the future solar availability and the future demand for energy.

There are long and short term variations in solar availability. The longest cyclic change is represented by a year. Storage systems attempting to cover such periods are referred to as seasonal stores. The shortest cyclic variation in solar energy is one day and this can be provided for by the use of diurnal storage systems.

Finally the design of the store and the materials used will depend on the temperature range over which the system operates.

4.1 Sensible Heat Storage

When charging or depleting a store results in the increase or decrease of the storage medium's temperature the system is referred to as a sensible heat store. In such systems the energy content is usually proportional to the temperature of the store. This is not always entirely true since the heat capacity of the storage medium may not always have a linear temperature dependence. However, such variations need concern only the specialist engineer and designer and temperature may be taken as a good indicator of the amount of energy stored.

Commonly available substances used for sensible heat storage are water, earth, stone, brick and concrete. Each has some properties which makes it better for some specific purpose than the others.

The most popular storage medium is water. It shows very good linearity of heat capacity with temperature over its practical operating range (10 – 90 °C). It also has the advantages of being non-toxic, cheap, widely available, and people are used to handling it.

When designing a heat store it is usually important to find a storage material which will store energy in a small volume. An important criterion is that the product of the specific heat capacity and the density of the material should be large. Water is particularly attractive in this respect since it has a higher value of this product than any other commonly available material. Water is also favoured because, being liquid over the practical operating temperature range, it is easily transported and thus can be used as a carrier of energy. It can be pumped and piped and its quantity can be measured and controlled.

Water has two main drawbacks: it causes corrosion and it supports organic life (algae). However, both problems can be overcome by careful selection of materials and proper design of the system. Small solar water heating installations, which today represent the majority of solar applications in the United Kingdom, almost invariably use copper cylinders as storage tanks and water as the storage medium. The heating and plumbing trades are used to working with these components and the corrosion or algae problems are minimal.

The size of the store affects the cost of the installation and an economic optimum has to be found taking account of available standard tank sizes. For domestic water heaters the British Standards Institution's code of practice (BS 5918, 1980) refers to the rule of thumb in sizing preheat vessels: 50 litres per square metre of collector area. Brinkworth (1980) identified store volume per collector area as a low sensitivity factor if the ratio is larger than 50. Work in Birmingham (UK) by Jesch (1978) and Soldatos indicated an economic optimum of 65 kg/m^2. For the US Lof (1977) suggests up to 80 kg/m^2, Beckman (1977) and co-workers used 75 kg/m^2. For interseasonal applications Braun (1980) pointed out the advantages of large ratios of 3-5000 kg/m^2 which can increase annual solar contribution from about 60% to 100% for a 50 m^2 installation in Madison, Wisconsin.

An interesting early application of interseasonal storage at the National Centre for Alternative Technology in Machynlleth, Wales, has been reported by Todd (1978). There an underground store with a storage mass to collector area ratio M/A of 1000 kg/m^2 was installed in 1977.

Horster (1980) used the Philips simulation model to analyse the M/A ratio in a series of studies for various Hamburg installations and in the recently published Solar Energy Technology Handbook edited by Dickinson (1980) and Cheremisinoff, there is a good analysis of low temperature sensible heat storage by Connor (1980). In addition to providing a survey of the subject he shows that a variety of different designs and applications across the continental United States tend to conform to an asymptotic value of 300 MJ/m^2K thermal storage heat capacity above which little gain can be expected in the fraction of the load which is met by solar radiation. This is equivalent to an M/A ratio of 72 if water is used as the storage medium.

An unusual building described by Williams (1978) is the Shenandoah Solar Recreation Center, not far from Atlanta, Georgia, is built partially underground. It has 1040 m^2 of flat plate collector augmented by 2140 m^2 of highly polished reflector. The storage consists of 65 m^3 of hot water and a 114 m^3 chilled water. These numbers reflect the fact that the cooling load is much larger at this location and although the building is well protected from excessive solar gains the air conditioning load still requires an M/A twice as large as that needed for heating. The combined heating, cooling and water heating load is 3.8×10^{12}J and 65% of this energy need is supplied by solar. Due to the underground structure, the heating load is only 2% of the total thermal energy load.

Figure 4.1.1: Solar domestic water heaters in Israel.

A storage tank may be heated directly by exposing it to the sun. Tanks without insulation as solar collectors have been used in Italy, the USA and in Israel, as shown in Figure 4.1.1. Recently Taha (1980) and Eldighidy in Saudi Arabia studied solar storage tanks with diathermanous insulation. The same concepts can be used in northern climates but with different surface to volume ratios. Such a collector-storage system, shown in Figure 4.1.2, was installed in the recently completed Superhome 2000 in Milton Keynes.

Figure 4.1.2: Solar collector-storage panels installed in the roof of this glass covered enclosure give off heat at night collected during the day.

4.2 Stratification

Large volume liquid heat stores often contained temperature gradients. These are caused by the warmer, less dense, fluid rising to the top of the tank whilst the colder denser fluid sinks to the bottom. This process of temperature stratification when it occurs improves the performance of the system and the designer should therefore try to exploit it. If the coldest water is drawn from the bottom of the tank and pumped to the collector inlet the collector will operate more efficiently than if the fluid at the top of the store is used. In a simple system stratification requires a careful compromise between the shape and size of the store, flow rates through the collector, demand flow rate and the positioning of the tank inlets and outlets.

Natural stratification is sometimes destroyed by vertical wall conductance. Several groups including Lavan (1977) and Thompson, and Sherman

(1979) and his co-workers, have investigated this phenomenon experimentally. The latter work was carried out in Arizona State University under the supervision of B. Wood and concluded that the optimum tank should be made of material which has a thermal wall conductance not greater than the conductance of the storage medium. They also recommend that the tank should be insulated more at the top than at the bottom. Turbulent mixing caused by badly designed inlet and outlet ports can also destroy stratification.

Natural stratification has attracted much research and invention in pursuit of practical solutions. An ingenious device reported by van Koppen (1974) which became well known and is shown in Figure 4.2.1. It consists of a vertical water tank with a flexible feed pipe on the side about mid height. This thin plastic tube has a negligible mass and floats to the exact position where the water returning from the collector should be deposited in the stratified tank. More recently van Koppen (1979) and co-workers have installed and monitored such devices in the Solar House in Eindhoven and in a government laboratory in Enschede. They reported that the floating inlet tube helped stratification and increased heat gain of the system by more than 5%.

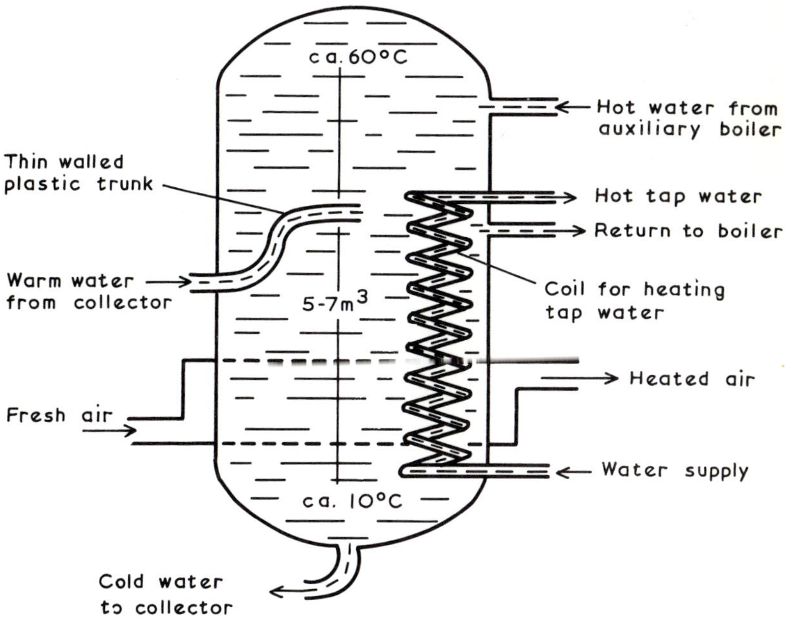

Figure 4.2.1: Natural stratification is helped by a floating thin walled plastic trunk which delivers water to the correct height as temperatures change.

A flexible floating inlet tube automatically makes the decision about the height at which the water has to be delivered in the tank to achieve good stratification. An alternative method is to force stratification by separating the water into tanks at discreet temperature levels. A solar water heating system using three separate tanks was reported by Jesch (1979) and Zekios in Birmingham. Jesch (1981 A) and Greeves analysed the effects of different combinations of tank sizes, tank insulation and weather patterns. Figure 4.2.2 shows a schematic layout of the different arrangements. A system is currently being monitored and early indications are that the forced stratification effect does extend the time periods when solar energy can be collected and increases the total solar contribution. The success of the scheme depends largely on the control strategy operated by an intelligent computer which not only follows the preprogrammed control strategy reported by Jesch (1981 B) and co-workers but also adapts it to the changing conditions.

Several mathematical analyses of stratified storage systems have been reported and the one by Duffie (1980) and Beckman presented a three node model. Veltkamp (1981) using a more detailed multinode model shows that the three node model is likely to underestimate the improvement in system performance caused by stratification. In practice however the complexity of the equipment and the associated cost often restrict the design of the system to simpler solutions which can be adequately represented by the three node model.

4.3 Rock bed storage

A packed bed of pebbles, rock or gravel of carefully selected sizes can be used as heat store. Air forced through the bed using a fan can exchange energy with the rock either by heating or cooling the store. The specific heat of rock is much less than that of water. Further about one third of the volume of a rock bed is void due to the packing of the rocks. This makes rock bed storage units larger and heavier than water stores of equivalent storage capacity. Yet there are advantages in the clean and simple application of this low technology solution. They do not suffer from freezing and leaks are less of a problem, although leaks, and the consequent loss of energy, are more difficult to detect. A further advantage of rock beds is the temperature stratification which can be established by the fan forced air flow, and which can be made use of by proper design.

An important analysis of rock bed stores used in solar systems was made by Balcomb (1975) and co-workers. The Argonne National Laboratory report (1979) describes rock beds in the range of 1 to 4 metres in length, filled with uniform size rock in the range of 30 to 120 mm diameter which will give reasonably good heat transfer without significant pressure drop. Mathematical models for the charging and discharging of rock beds are described in many publications, for example the one by Sowell (1980) and

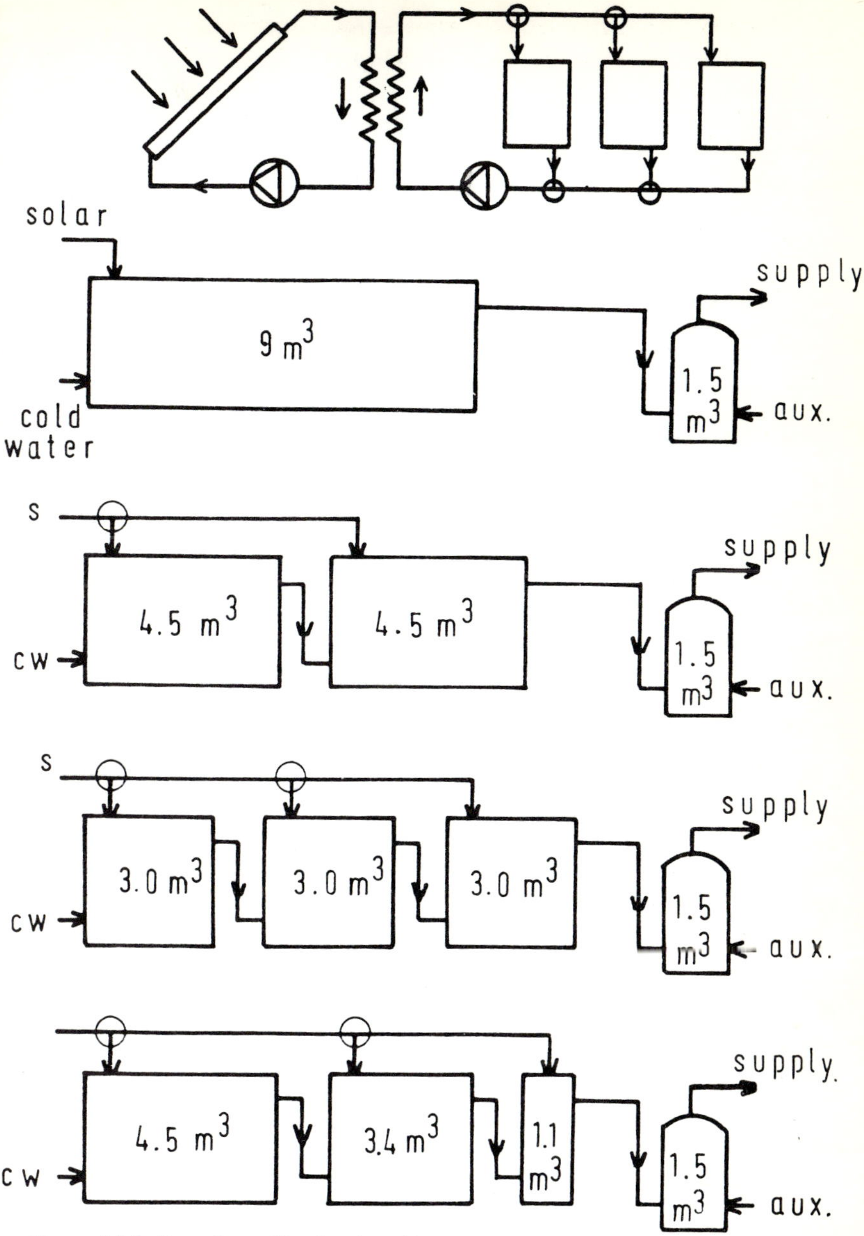

Figure 4.2.2. Forced stratification is achieved by three port valves chanelling the indirectly heated water to the correct tank as temperatures change. More separate tanks of the same total store capacity give higher system efficiency.

Curry. In Wisconsin, Persons (1980) and co-workers used the TRNSYS model to predict rock bed performance which was subsequently verified by experiments. The agreement was within 5%.

4.4 Latent heat and chemical storage

One disadvantage of sensible heat stores is that their temperature varies with energy content. This can adversely affect the efficiency of solar energy collection, storage and delivery. The search for other methods of storing thermal energy has been on for a long time.

Some materials have a large latent heat capacity when changing phase while their temperature remains almost constant. This means that unusually large amounts of energy can be transferred into them over a narrow temperature range in which the change of phase occurs. For example the melting of a salt. Later the reverse process can be used to remove energy from the molten salt by cooling it and allowing it to solidify. Chemical reactions are also associated with thermal processes, that is, the material absorbs or releases large amounts of energy when they react. In selected cases the chemical change is reversible and the charging and discharging of a chemical store can be carried out over a small temperature range in a manner analogous to latent heat storage. Because the latent heat and the heat of reaction increase the thermal capacity of the store over a given temperature range, use can be made of these phenomena to reduce the size of thermal stores.

The early pioneering work in this field was done by Telkes (1949) who developed a method of charging and discharging stores, filled with sodium sulphate decahydrate and its eutectics, more than a thousand times in a continuous experiment. Further progress was reported by Telkes (1977) in finding the right additives to prevent segregation of the anhydrous components of higher density thus increasing the number of useful cycles of charge and discharge. Recently in the UK, Page (1981) and Swayne reported that a latent heat store has been introduced onto the market and installed into new solar houses offered for sale on the open market.

One of the difficulties in latent heat systems is the fact that the heat transfer coefficient of the solid phase is often very low. Therefore the rate at which energy can be transferred varies with the condition of the storage material. In practice the difficulty can be reduced by increasing the heat exchange surface area. This can be done by using fins, as suggested by de Jong (1981) and Hoogendoorn or by encapsulating the latent heat material as described by Mehalik (1975 and 1977) and Tweedie or by using a direct contact heat transfer system as reported by Fouda (1980) and co-workers. An alternative developed by Thornton (1979) and Herrick prevents the solid phase from adhering to the heat transfer surface. Marshall (1978 and 1981) has reported initial results on this subject using a paraffin wax.

A Japanese team working on high temperature applications of solar heat storage modelled nitrate salts and verify the model with experiments in the 200-300°C range, but some of the difficulties of quickly changing heat transfer coefficients have yet to be resolved according to Kamimoto (1980) and co-workers.

A German group reported experimental results with a system using a zeolite-water system for low temperature storage at high energy densities which can be used to drive a chemical heat pump. The authors, Khelifa (1980) and co-workers, are optimistic about doubling the energy density in the near future.

In Yugoslavia, Kulisic (1980) and co-workers are working on heat-of-solution thermal energy storage and in Sweden Carlsson (1980) and co-workers, supported by the Swedish Council for Building Research, are developing a heat-of-fusion short term solar energy storage unit. Another Japanese team Yoneda (1980) and co-workers is using a pilot plant to evaluate various materials for use in chemical reactions which provide long term solar energy storage. Busico (1980) and co-workers in Italy are working on solid-solid phase transition compounds. Work at the Rutherford Laboratory on chemical heat pumping and energy storage using the reaction between sulphuric acid and water has been reported briefly by Smith (1978). It is obvious that all over the world attempts are being made to find new ways of storing low grade energy.

4.5 **Solar ponds**

Salt rich water lakes under high altitude solar radiation can develop a temperature distribution in the body of the water whereby the top is colder than the bottom. This is due to two main reasons: (a) the short and medium wavelength parts of the spectrum penetrate deep into the water heating the bottom, while only a small portion of the solar radiation, that with longer wavelengths, is absorbed nearer the surface, (b) the salt concentration gradient prevents convection because a highly concentrated solution of salt, even when hot, is denser than a cool weak solution. As a consequence the lake becomes a solar trap with the warmer water at the bottom of the pond.

The salt concentration has a positive gradient with depth partly because of natural dilution near the surface and partly because of sedimentation. This salt gradient is vitally important in establishing the temperature profile which is also positive with depth. Any movements, natural or man-made, will upset the salt gradient and will convect heat from the bottom of the lake. The removal of useful heat stored in such solar ponds creates various problems, some connected with disturbing the stratification of the water.

The pioneering work on solar ponds was done by Tabor (1963) in Israel using a 25 m x 25 m by 0.8 m deep pond. Later other ponds were built in

Israel, India, Canada and the United States. The filling of artificial solar ponds is a slow and delicate operation since it is necessary to establish the salt gradient from the very beginning and not cause convective currents at the bottom. Plastic liners are used to prevent the loss of salt and water to the soil and in some instances covers have been tried in order to reduce losses from the surface of the pond. Continuous maintenance of salt concentration requires regular topping up with salt to compensate for the losses by diffusion.

After filling the pond it takes several years to reach an equilibrium condition. A study, by Akbarzadeh (1980) and Ahmadi completed recently in Iran concluded that heat should not be removed from the pond before it reached a state of full charge but afterwards heat should be removed continuously to keep the energy balance of the pond at a level consistent with the expected availability of solar radiation.

It should be mentioned that most authors prefer to use the term 'salt gradient solar ponds' in order to differentiate them from 'shallow solar ponds' which are discussed below. It is interesting to note that salt gradients are not the only mechanisms for trapping solar energy under water. Hull (1979) reported membrane stratified solar ponds in which suitable geometries of the convection suppressing membranes help to achieve mush deeper heat storage layers than are possible with salt solutions.

Another effort to optimise the solar pond was described by Nielsen (1979) a leading practitioner in the United States. He proposed a continuous liquid injection procedure to modify the salinity profile, and keep the surface zone thin, so that more energy could pass through it and thus optimise the gradient zone.

Saturated salt solar ponds are being studied by Ochs (1979) and Bradley, and Kooi (1981). The solubility of the salt used increases such that the density of the saturated also increases with temperature hence the temperature is naturally maintained.

Shallow solar ponds were developed from the techniques used in solar stills. The solar still evaporates water from a saline solution contained in a tray which is only a few centimetres deep. Shallow solar ponds have a similar depth of water, but the cooling effect of the evaporation is prevented by a transparent film on top of the water. Sometimes a second film separated from the first by an air gap is placed over the top to act as a double glazing. Clark (1980) and Dickinson at the Lawrence Livermore Laboratory in California have tested a number of designs, most of them aimed at industrial or agricultural use.

It is interesting to note that Carter (1979) in the south of England experimented in the late 1950's with the idea of shallow solar ponds for swimming pool heating. Her experiments included studying the effect of the absorptivity of the bottom of the shallow pond but did not include the control of evaporation.

The control of evaporation is a central theme in solar pond research of every kind. As Duffie (1980) and Beckman say about solar ponds: "They are all horizontal collectors, some of which have integral storage, and the evaporation of water is either an objective or a problem".

Finally a note on the history of solar ponds which sums up the ways in which the process of invention works. It was first noticed in Hungary by Kalecsinsky (1902) at the beginning of the century that lakes with high salt contents can act as 'heat accumulators' of solar energy. His studies did not lead to a practical conclusion until Bloch in Israel made a proposal in 1948 which was followed by research and the first experimental pond was created in 1959. Tabor (1963) wrote an article in the Solar Energy Journal, reporting on the project and again in 1965 with Matz. Work continued until 1967. Various difficulties halted progress until 1975, when they were eventually overcome and at present Bronicki and his company Ormat Turbines Ltd. together with another firm Solmat Systems Ltd. are co-operating with nine Israeli universities and engineering institutions. They have built a 150 kW turbine plant at Ein Bokek on the shores of the Dead Sea, Figure 4.5.1, which is operating successfully and another power plant with 5000 kW power output which will be operational in 1982. This will be based on seasonal storage due to the size and design of the new pond.

Figure 4.5.1: Solar pond at Ein Bokek. Turbulence suppressors float on the surface. The turbines are to the left, the Dead Sea is to the right of the pond.

So in eighty years the accidental, though methodical, observations of one man in Hungary became the practical reality of a nation in Israel, serving the daily needs of the community.

References

Akbarzadeh, A. (1980) and G. Ahmadi: "Computer simulation of the performance of a solar pond in the southern part of Iran." **Solar Energy 24** 1980, 143-151.

Argonne National Laboratory (1979): **Design and Installation Manual for Thermal Energy Storage.** Solar Energy Group Report ANL-79-15, Feb. 1979.

Balcomb, J. D. (1975), J. C. Hedstrom and B. T. Rodgers: "Design considerations of air cooled collector/rock bin storage solar heating systems." ISES Congress, Los Angeles 1975, Paper 41/4, 314-315.

Beckman, W. A. (1977), J. A. Duffie and S. A. Klein: "Simulation of solar heating systems." in **Applications of Solar Energy for Heating and Cooling of Buildings.** ed. R. C. Jordan and Y. H. Liu. ASHRAE GRP 170, 1977, IX 1-9.

Braun, J. (1980): **Seasonal Storage of Energy in Solar Heating.** M.Sc. Thesis, University of Wisconsin, Madison, 1980.

Brinkworth, B. J. (1980): "The British Standard Code of Practice and its implications for solar hot water system design." in **Solar Energy in the 80s.** Conference (M2) Birmingham, UK-ISES, Midlands Branch, 1980, 79-84.

BSI 5918: Code of practice for solar heating systems for domestic hot water. British Standards Institution, London, BSI5918:1980.

Busico, V. (1980), C. Carfagna, V. Salerno, M. Vacatello and T. Fittipaldi: "The layer perovskites as thermal energy storage systems." **Solar Energy 24** 1980, 575-579.

Carlsson, B. (1980), A. Schmidt, A. Györki, H. Stymne and G. Wettermark: "Experimental characterization of the thermal performance of a heat-of fusion storage unit," in **Tagungsbericht des 3. Internationalen Sonnenforums-Freie Vortrage** -DGS-Sonnenenergie Verlag, Hamburg 1980, 658-665.

bericht des 3. Internationalen Sonnenforums - Freie Vorträge - DGS-Sonnenenergie Verlag, Hamburg 1980, 658-665.

Carter, Brenda (1979): Private Communication.

Clark, A. F. (1980) and W. C. Dickinson: "Shallow solar ponds." in **Solar Energy Technology Handbook, Part A Engineering Fundamentals.** ed. W. C. Dickinson and P. N. Cheremisinoff, Marcel Dekker 1980, 721-763.

377-402.

Connor, D. W. (1980): "Low temperature sensible heat storage." in **Solar Energy Technology Handbook. Part A Engineering Fundamentals.** ed. W. C. Dickinson and P. N. Cheremisinoff, Marcel Dekker 1980, 721-763.

de Jong, A. C. (1981) and C. J. Hoogendoorn: "Improvement of heat transport in paraffins for latent storage system." in **Thermal Storage of Solar Energy.** ed. C. den Ouden, Martinus Nijhoff, Hague, 1981, 123-133.

Duffie, J. A. (1980) and W. A. Beckman: **Solar Engineering of Thermal Processes.** Wiley-Interscience 1980, 762 p.

Fouda, A. E. (1980), G. J. G. Despault, J. B. Taylor and C. E. Capes: "Solar storage systems using salt hvdrate latent heat and direct contact heat exchanger — I. Preliminary design considerations." **Solar Energy 25** 1980, 437-444.

Hörster, H. (1980): **Wege zum energiesparenden Wohnhaus.** Philips Fachbücher, Hamburg 1980.

Hull, J. R. (1979): "Membrane stratified solar ponds." in **Sun II.** Proc. ISES Silver Jubilee Congress in Atlanta, 1979, Vol. 2, 1000-1004.

Jesch, L. F. (1978) and P. G. Soldatos: **Solar Water Heating Performance Analysis by Mathematical Modelling.** Research Report No. 166. University of Birmingham, Mech. Eng. Dept. 1978.

Jesch, L. F. (1979) and A. Zekios: "Adaptive control for a multiple tank, multiple temperature solar energy storage system." in **Sun II.** Proc. ISES Silver Jubilee Congress in Atlanta 1979, Pergamon Press, Vol. 1 852-856.

Jesch, L. F. (1981) and T. W. Greeves: "Solar energy storage in forced stratification tanks." Papers presented at the **International Conference on Energy Storage,** Brighton 1981, BHRA Fluid Engineering, Vol. 1, 247-257.

Jesch, L. F. (1981), T. W. Greeves and D. Haralambopoulos: "Simulation of optimal control strategies for large solar water preheat systems." Proc. ISES Solar World Forum in Brighton 1981.

Kalecsinsky, A. V. (1902): "Uber die Ungarischen warmen und heissen Kochsalzseen als natürlich Wärmeaccumulatoren." **Ann. Physik 7,** (4) 1902, 408-416.

Kamimoto, M. (1980), T. Tanaka, T. Tnai and T. Horigome: "Investigation of nitrate salts for solar latent heat storage." **Solar Energy 24** 1980, 581-587.

Khelifa, N. (1980), D. Jung and R. Sizmann: "Zeolithe als langzeitspeicher solarer Energie zur Wärmeversorgung im Niedertemperaturbereich." **Tagungsbericht des 3. Internationalen Sonnenforums - Freie Vorträge -** Hamburg 1980, DGS-Sonnenenergie Verlag, 674-682.

Kooi, C. F. (1981): "Salt gradient solar pond with reflective bottom: application to the 'saturated' pond." **Solar Energy 26** 1981, 113-120.

Kulisic, P. (1980), J. Vulatin, J. Mastelic, M. Bosanac and B. Prib: "Thermal energy storage, using heat of solution." in **Tagungsbericht des 3. Internationalen Sonnenforums - Freie Vorträge -** Hamburg 1980, DGS-Sonnenenergie Verlag, 666-673.

Lavan, Z. (1977) and T. Thompson: "Experimental study of thermally stratified hot water storage tanks." **Solar Energy 19** 1977, 519-524.

Löf, G. O. G. (1977): "Systems for space heating with solar energy." in **Applications of Solar Energy for Heating and Cooling of Buildings.** ed. R. C. Jordan and Y. H. Liu, ASHRE GRP 170, 1977, XII 1-19.

Marshall, R. (1978): "Studies of natural convection effects in an annulus containing phase change material." in **Storage in Solar Energy Systems.** Conference (C15) London, UK-ISES 1978, 11-30.

Marshall, R. (1981): "Experimental experience with the ASHRAE/NBS procedures for testing a phase change thermal storage device." Papers presented at the **International Conference on Energy Storage.** Brighton 1981, BHRA Fluid Engineering, Vol. 1, 129-143.

Mehalik, E. M. (1975) and A. T. Tweedie: **Two Component Energy Storage Material.** General Electric Co., Space Division, Philadelphia, Report NSF/RANN/SE AER 74-09186, Nov. 1975.

Mehalik, E. M. (1977) and A. T. Tweedie: **Two Component Energy Storage Material Study – Phase Two.** Report COO/2845-77/1 March 1977.

Nielsen, C. E. (1979): "Control of gradient zone boundaries." in **Sun II.** Proc. ISES Silver Jubilee Congress in Atlanta 1979, Pergamon Press, Vol. 2, 1010-1014.

Ochs, T. L. (1979) and J. O. Bradley: "The physics of a saturated $NA_2O,2B_2O_3.10H_2O$ non-convecting solar pond." in **Sun II.** Proc. ISES Silver Jubilee Congress in Atlanta 1979, Pergamon Press, Vol. 2, 1026-1028.

Page, J. K. R. (1981) and R. E. H. Swayne: "Phase change thermal storage for solar applications." Papers presented at the **International Conference on Energy Storage.** Brighton 1981, BHRA Fluid Engineering, Vol. 1, 165-170.

Persons, R. W. (1980), J. A. Duffie and J. W. Mitchell: "Comparison of measured and predicted rock bed storage performance." **Solar Energy 24,** 1980, 199-201.

Sherman, C. (1979), D. D. Wood and J. Mason: "Effect of vertical wall conductance on temperature relaxation in thermally stratified liquid thermal storage tanks." in **Sun II.** Proc. ISES Silver Jubilee Congress in Atlanta 1979, Pergamon Press, Vol. 1, 591-595.

Smith, P. F. (1978): "Chemical heat pumping and energy storage." in **Storage in Solar Energy Systems.** Conference (C15) London, UK-ISES 1978, 55-58.

Sowell, E. F. (1980) and R. L. Curry: "A convolution model of rock bed thermal storage units." **Solar Energy 24,** 1980, 441-449.

Tabor, H. (1963): "Large-area solar collectors for power production." **Solar Energy 7,** 1963, 189-194.

Tabor, H. (1965) and R. Matz: "A status report on a solar pond project." **Solar Energy 9,** 1965, 177-182.

Taha, I. S. (1980) and S. M. Eldighidy: "Analysis of solar storage tanks, having diathermanous insulation." in the **Solar Energy in the 80s.** Conference (M2) Birmingham, UK-ISES Midlands Branch, 1980, 85-91.

Telkes, M. (1949): "Storing solar heat in chemicals." **Heating and Ventilating 46,** Nov. 1949, 79-86.

Telkes, M. (1977): "Solar energy storage." in **Applications of Solar Energy for Heating and Cooling of Buildings.** ed. R. C. Jordan and Y. H. Liu. ASHRAE GRP 170, 1977, VII 1-9.

Thornton, R. F. (1979) and C. S. Herrick: **Thermal Energy Storage Sub-Systems for Solar Heating and Cooling Applications (Rolling Cylinder Thermal Storage).** General Electric Co., Schenectady, Report SRD-79-089, June 1979.

Todd, R. W. (1978): "A solar heating system with interseasonal storage." in **Storage in Solar Energy Systems.** Conference (C15) London, UK-ISES 1978, 1-10.

van Koppen, C. W. J. (1974): "Zonne-Energie Voor Huisverwarming en Warmwatervoorziening." Rapport WPS3-74.11.R222. Eindhoven Technological University, 1974.

van Koppen, C. W. J. (1979), J. P. Simon Thomas and W. B. Veltkamp: "The actual benefits of thermally stratified storage in a small and medium size solar system." in **Sun II,** Proc. ISES Silver Jubilee Congress in Atlanta 1979, Pergamon Press, Vol. 1, 576-580.

Veltkamp, W. B. (1981): "Thermal stratification in heat storage." in **Thermal Storage of Solar Energy.** ed. C. den Ouden, Martinos Nijhoff, Hague, 1981, 47-59.

Williams, J. R. (1978): "A completely instrumental solar building: the Shenandoah Solar Recreational Center." in **Solar Energy Conversion.** (5th Course, Univ. of Waterloo, Canada, 1978). ed. A. E. Dixon and J. D. Leslie, Pergamon Press, 1979, 417-433.

Yoneda, N. (1980), S. Ito and S. Hagiwara: "Study of energy storage for long time, using chemical reactions." in **Tagungsbericht des 3. Internationalen Sonnenforums - Freie Vorträge -** Hamburg 1980, DGS-Sonnenenergie Verlag, 648-657.

Chapter 5

PHOTOVOLTAIC SOLAR CELLS

Photovoltaic solar cells convert daylight into direct-current electricity silently, cleanly and with no harmful waste products. Figure 5.0.1 shows a typical example. It is 100 mm square by about ⅓ mm thick and is made of a crystalline form of silicon, one of the most common and available elements on earth. In bright sunlight, under so-called ''peak'' conditions, that is at an intensity of 1000 W/m^2 and a temperature of 25°C, it will produce over 1 W at just under 0.5 V. However, it does not need direct sunlight and will operate efficiently in overcast conditions, with little reduction in voltage although the current produced is proportional to the irradiance. Moreover, it is more efficient at low temperatures and responds very rapidly to sudden changes in irradiance. These properties make solar cells particularly attractive for places like NW Europe, where over half the annual solar input is diffuse.

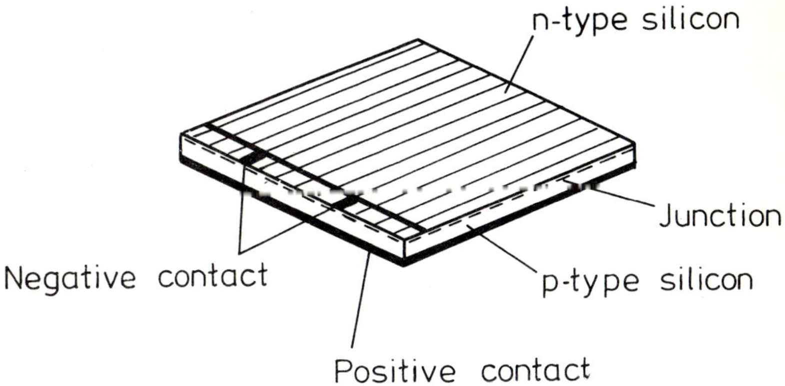

Figure 5.0.1: Photovoltaic solar cell with silicon p-u junction.

Because of the variability of the conditions of irradiance and temperature under which terrestrial photovoltaic generators have to operate, it is usual to rate their output in terms of "peak" power, i.e. the maximum power produced by the generator under "peak" conditions. When comparing photovoltaic generation with conventional electricity generation, it is important to remember that the average output power will be less than its peak power rating because the average irradiance is less than 1000 W/m^2.

With solar cells, one can construct generators of any size and voltage from mass-produced modules and, most importantly, locate them where the power is needed, with consequent saving in transmission costs. It is a mistake to think of photovoltaic generators only in terms of large central power stations. As they have no moving parts to wear out, they can operate reliably for many years with little or no maintenance, provided they are properly engineered to withstand the weather.

It was Becquerel (1839) who first observed the "photovoltaic effect" when he directed light on to one of the electrodes in an electrolyte. Adams (1877) and Day noted the effect in selenium some 38 years later and subsequent work by Lange (1930), Grondall (1933) and Schottky (1930) on selenium and cuprous oxide cells led to the development of the photographic exposure meter and many other photocell applications. But it was not until 1954 that solar cells with a sufficiently high conversion efficiency for power generation came on the scene. In that year, Chapin (1954) and co-workers reported that they had made silicon solar cells with an efficiency of 6% and Reynolds (1954) and co-workers made a similar breakthrough with a cadmium sulphide device.

Silicon solar cells soon found an application in the space programme and developed rapidly in conversion efficiency and reliability. It is safe to say that the exploration and exploitation of space could not have developed in the way they have, had it not been for the timely invention of this long-term power generating device. Over the past two decades, photovoltaic power generation in space has become a well-established technology with proven reliability.

Since the 1973 fuel crisis, interest in solar cells for terrestrial power generation has increased enormously and photovoltaic conversion is now a significant element in most renewable energy R & D programmes, particularly those of USA, Japan and the CEC. Although still high, the capital cost has been brought down by an order of magnitude compared with space systems and the downward trend is expected to continue, helped by the R & D effort and a rapidly expanding market. Annual production of silicon cells in USA is expected to rise from 4 MWp (4 MW peak) in 1980 to 7 MWp in 1981 and 14 MWp in 1982. Japan's targets are 50 MWp/year by 1985 and 3 GWp/year by 1990. The CEC has recently announced new development plans. Accordingly, by the year 2000, installed photovoltaic generating capacity in Europe should reach 10 GWp. In the European climate this is equivalent to a continuous generation capacity of 1 GW or 0.5% of the total installed European generating capacity. There are strong indications that photovoltaics will become a major new industry in the course of the next decade.

A good description of the principles of photovoltaic conversion is given by Backus (1980) where the reader can find information also on the mathematical modelling of solar cells.

5.1 **The physics of silicon solar cell**

Referring to Figure 5.0.1 again, a silicon cell is made from a thin wafer of crystalline solution, doped with boron to a resistivity of 1 to 10 ohm.cm. Phosphorus is diffused into the water to form a p-n junction a fraction of a micron below the active surface. The front (negative) contact is in the form of a narrow-fingered grid, while the positive contact usually covers the entire back surface. The front is treated with an antireflective coating.

The mode of operation and performance characteristics of these cells are fully described in the literature such as by Prince (1955), Rappaport (1959), and Wysocki (1960) and Rappaport, but here is a brief explanation.

The phosphorus causes an excess of conduction-band electrons in the silicon and the boron an excess of holes (valence electron vacancies), which act like positive charges. At the junction, conduction electrons from the n-region diffuse into the p-region and combine with holes, cancelling their charges. The opposite action also takes place, holes from the p-region crossing into the n-region and combining with electrons. This results in the disappearance of electrons and holes from the vicinity of the junction (the so-called "depletion area" or "barrier layer") and leaves behind layers of charged impurity atoms, which are positive in the n-region and negative in the p-region. Thus a reverse electric field is set up around the junction.

When light falls on the active surface, photons with energy exceeding the bandgap or energy gap E of the semiconductor (i.e. the potential difference between the top of the valance band and the bottom of the conduction band, – 1.1 eV in the case of silicon) interact with the valence electrons and elevate them to the conduction band. This movement leaves behind holes, so the photons are said to generate "electron-hole pairs". In crystalline silicon, sunlight generates these pairs throughout the thickness of the cell, in concentrations depending on the intensity and spectral composition of the light. The electrons and holes diffuse through the crystal, the electrons moving directly and the holes more slowly by valence electron substitution from atom to atom. Some recombine, neutralising their charges and giving up energy in the form of heat. Others reach the junction and are separated by the reverse field, the electrons being accelerated towards the negative contact and the holes towards the positive. If the cell is connected to a load, there will be a drift of electrons from the negative contact through the load to the positive contact. In silicon, the generated current in any particular type of light is directly proportional to the irradiance and the active area of the cell.

The generated current is composed of increments produced by photons of various wavelengths. If the incremental current generated by unit irradiance is plotted as a function of wavelength, the resulting curve is called the "absolute spectral response". Figure 5.1.1 shows a typical example. Note that it covers the whole of the visible spectrum and part of the infrared, peaking at about 0.9 microns.

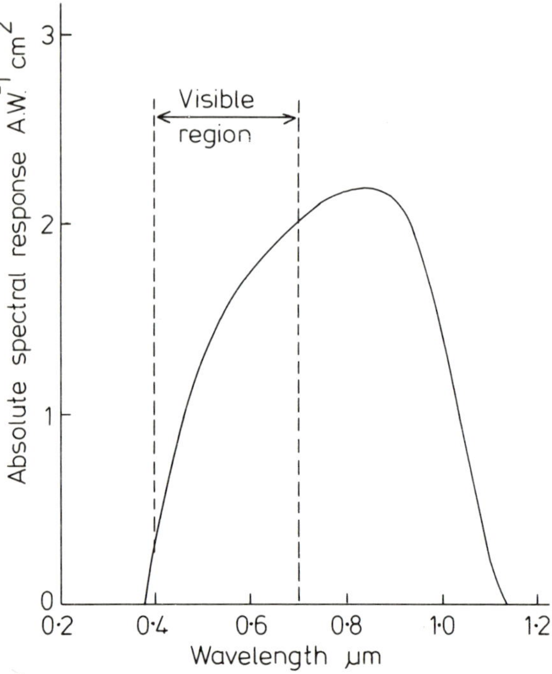

Figure 5.1.1: Absolute spectral response versus wavelength of a silicon solar cell.

Figure 5.1.2 shows the voltage current characteristics of a modern silicon solar cell at 25°C in sunlight at an irradiance of 1000 W/m^2. Maximum power is produced at a voltage between 0.4 and 0.5 V, depending on the base resistivity and quality of the cell. The conversion efficiency is the maximum output power expressed as a percentage of the input power which is the product of irradiance and cell area. In the best terrestrial cells, it is 12-13% under peak conditions. The operating voltage is normally fixed so as to be slightly below the voltage for maximum power at the operating temperature.

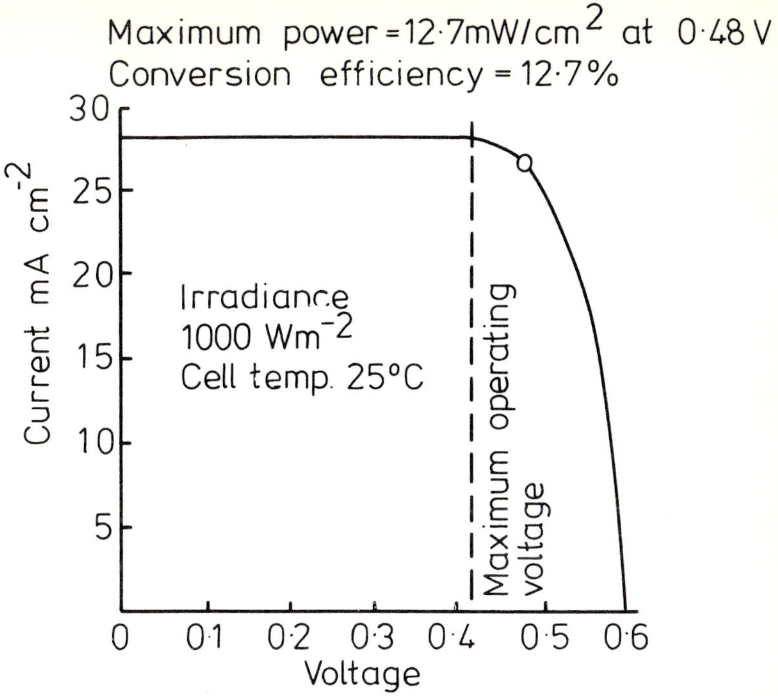

Figure 5.1.2: Current-voltage characteristics of a silicon solar cell operating under stated conditions.

Changes of irradiance shown in Figure 5.1.3 affect the short-circuit current proportionally but have little effect on the open-circuit voltage until very low levels are reached, because the latter is logarithmically related to the irradiance. In concentrated sunlight, the short-circuit current is proportional to irradiance up to extremely high levels, provided the temperature is kept constant. As the open-circuit voltage also increases, one would expect the conversion efficiency to improve and so it does, up to a point. But the internal resistance of the cell has a progressively flattening effect on the characteristic and this limits the improvement that can be achieved.

An increase in temperature causes a slight rise in short-circuit current but a sharp fall in open-circuit voltage and maximum power as illustrated in Figure 5.1.4. Typically, the power output falls about 0.5% for every °C rise in temperature, so it is important to ensure that the cells run as coolly as possible under operational conditions. Note that the voltage for maximum power also falls with increasing temperature but, provided the cell operating voltage does not exceed the voltage for maximum power at the highest operational temperature, the output current is little affected by changes of temperature below the maximum.

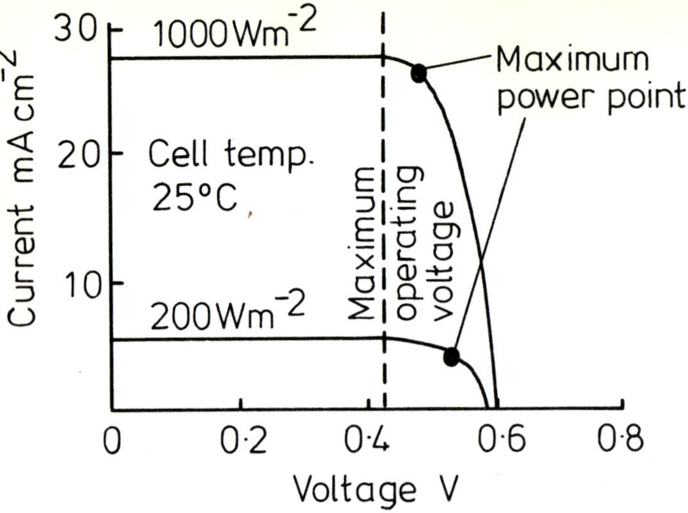

Figure 5.1.3: Changes of irradiance affect current proportionally but voltage is not affected much.

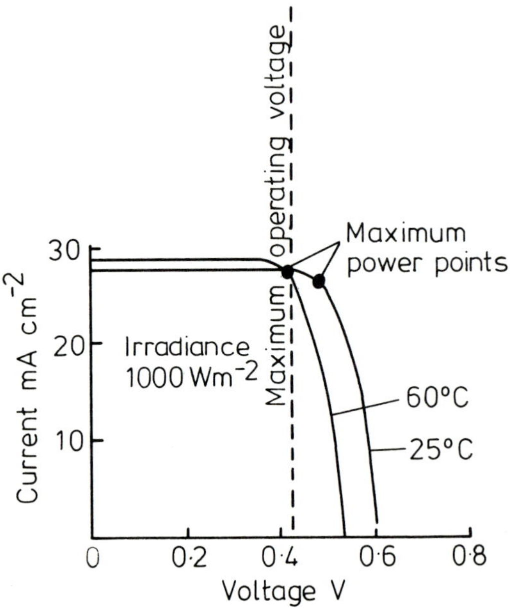

Figure 5.1.4: Temperature changes have little effect on the current but cause a sharp change in the voltage.

5.2 The Photovoltaic Module

The module is the basic building block of a photovoltaic array. It consists of a number of interconnected cells, usually but not always in a single series string, encapsulated behind a transparent window for protection against the weather. It is bolted or clamped to a supporting framework and has terminals or a pigtail lead for connection to other modules or the load. The most common window material is tempered glass and the encapsulant is either a transparent silicone rubber, polyvinylbutyral (pvb) or ethylene vinyl acetate (eva). The current target in module engineering is a lifetime of 20 years. Considerable progress has been made towards this goal by applying the lessons of field trials reported by Shumka (1978) and Stern as well as Anagnostou (1978) and Forestieri and most makers now give a 5-year guarantee. Authorities in the USA and Europe have prepared specifications of standard laboratory tests such as by JPL (1981) and CEC (1981) to check the ability of modules to withstand many years of UV radiation, thermal cycling, humidity, rain, hail, winds, snow, ice.

5.3 Storage

In many applications, it is necessary to store some of the generated electricity for use during periods of darkness, low insolation and peak loads. Although many new storage techniques are being developed and it may be economic under favourable conditions to use hydraulic storage, the practical solution for most applications is the electrochemical battery.

Batteries for photovoltaic systems need to have good charging efficiency low open-circuit losses and a long trouble-free life under difficult cycling and environmental conditions. No existing battery fulfils all these requirements at low cost but some R & D work in this area is being sponsored by the EEC and the US Department of Energy. In the meantime, lead-acid batteries and, to a lesser extent, nickel cadmium are being used.

A review of the current status and future plans of developing advanced batteries for energy storage was given recently by Jensen (1981) and Tofield. They also described the current Anglo-Danish programme partially funded by the CEC which is aimed at developing new materials.

5.4 Present Applications

Despite their high capital cost, solar cells are already competitive on a life-cycle cost basis in a wide range of low-power applications, such as:

- Remote marine and aircraft beacons, navigation lights, fog signals.
- Remote telecommunication links, automatic weather stations, highway breakdown telephones and warning lights.
- Cathodic protection for pipelines, wellheads, bridges, etc.
- Battery charging for remote homes, camping, boats, and to replace dry batteries in calculators, watches, etc.
- Mobile, silent electrical power supplies for military equipment.

space

As an example of such applications the photograph Figure 5.4.1 shows a solar-powered navigation beacon which was installed in Chichester Harbour by the Royal Aircraft Establishment in May, 1977, and has operated satisfactorily since.

Figure 5.4.1: Solar powered navigation beacon in Chichester Harbour.

A typical stand-alone system consists of an array of flat-plate modules facing the Equator at a fixed inclination, charging batteries which supply the load directly or through power conditioning equipment. A simple electronic charge regulator is included when necessary to protect the batteries against overcharge during periods of low usage or after a succession of long, sunny days. Solid-state diodes protect the battery from short circuits in the array and prevent the solar cells loading the battery when they are not illuminated.

The system designer has to work out the cheapest combination of array area, inclination and storage capacity that will meet his load requirements under the worst conditions at the selected site throughout the demand period. Of course, site location affects the size of array needed for any particular job but not as much as might be expected. As a rough guide, Table 5.4.1 gives the module area required to yield a mean output of 2.4 kWh/day (i.e. 100 W continuous) in various parts of the world. The calculations are based on annual mean global irradiance data by Budyko (1958), a module conversion efficiency of 10%, an allowance of 25% for battery losses and a 5% allowance for other losses, such as window contamination. At an efficiency of 10%, 1 m^2 of modules is equivalent to 100 Wp.

Table 5.4.1. Approximate Array Sizes for a Mean Output of 2.4 kWh/day

Region	Annual Mean Global Irradiance W/m^2	Minimum Module Area m^2
NW Europe	110	12.8
S France S Germany Japan N USA New Zealand	150	9.4
Brazil East Indies Greece Italy Spain S Australia Mid & SE USA	200	7.0
California India Middle East Mid Australia S Africa Sahara West Indies	250	5.6
Red Sea	300	4.7

Note that an array in NW Europe would have to be almost three times the size of one performing the same duty in the sunniest part of the world, the Red Sea area. The problem in the higher latitudes is not so much the total available solar energy as the marked seasonal variation.

In some cases a continuous load is not required and 100% availability may not be needed. These factors and the relative costs of batteries and arrays may result in optimal designs different from those given in Table 5.4.1. Methods of estimating size requirements for photovoltaic systems have been presented by Baikie (1978), and by Darlington (1978) and co-workers.

5.5 Cost Reduction Programmes

In the United States in 1975 the price of a module, the biggest single cost element in a photovoltaic system, ranged from \$35 to \$40/Wp. In 1981, modules with far better reliability and service life, are selling there commercially for about \$10/Wp, although lower prices have been bid for recent Government bulk purchases. In Europe, prices are somewhat higher. "Balance-of-systems" costs, which include such items as the array structure, wiring, power conditioning, energy storage, marketing, transportation and installation, can make the total system price from 1.5 to 5 times the module price, depending on cell efficiency, location and the complexity of the system.

Table 5.5.1 shows the United States Department of Energy (US DOE) current price goals in 1980 dollars, with corresponding prospective applications. Many regard these targets as optimistic and they certainly are for Europe. Nevertheless, there is considerable confidence that the quoted price levels will be achieved, even though the dates may slip.

Table 5.5.1. US DOE Price Goals

Year	Module Price \$/Wp	System Price \$/Wp	Prospective Market
1982	2.80	6.00 – 13.00	Small, remote loads
1986	0.70	1.60 – 2.20	Utility-connected residences and intermediate load centres
1990	0.15 – 0.50	1.10 – 1.30	Central power stations

A substantial part of the current R & D effort is devoted to finding cheaper and more energy-efficient ways of producing the silicon crystal wafer, which at present accounts for about 20% of the module cost.

The present Siemens process for producing the high purity semiconductor-grade polysilicon feedstock for crystal growing consumes about

2000 kWh/kg and costs about $63/kg. Union Carbide, Hemlock and Battelle are working on different processes which should reduce the energy consumption to about 90 kWh/kg and the cost to between $12 and $20/kg as reported by Lutwack (1980). These are projected figures, based on plants producing 1000 tonnes/year. The U.S. Department of Energy price goal is $14/kg. Union Carbide are furthest ahead, with a 100 tonnes/year pilot plant scheduled to be in operation in December, 1981.

Dow Corning and Westinghouse are developing processes for producing low-cost "solar grade" silicon, which would not be as pure as the semiconductor grade but might be good enough for solar cells according to Hunt (1979) and Dosaj, and Reed (1978) and co-workers.

The present Czochralski crystal growing process can be cheapened by drawing several large ingots from a single crucible with melt replenishment. One firm, Kayex Corp., have succeeded in growing four 120 mm diameter ingots in this way and have made cells of about 10% efficiency from them. This was described by Kachare (1980) and co-workers during the 14th IEEE Photovoltaic specialists conference recently.

A cheaper alternative to Czochralski single crystal, which is now coming on to the market under the names of "Silso" and "Semix", is cast multicrystalline (or "semicrystalline") silicon. The square wafers have a better packing density than the round or semicircular cells made from Czochralski crystal and the energy consumed in their manufacture is an order of magnitude less – about 10 kWh/kg. Wacker Chemitronic, the manufacturers of Silso, have decided to invest heavily in this material and increase their production capacity in Europe to 1800 tonnes/year by the end of 1981 according to Freiesleben (1980). Similar plans have been revealed by the manufacturers of Semix as reported by Lindmayer (1980). Both firms are relying on the multi-crystalline material to overcome the present bottleneck in silicon supplies and meet the rapidly increasing demand for solar cells.

Another interesting approach was described by Schmid (1980) and co-workers. This is the directional solidification of silicon ingots by the heat exchange method (HEM). It is still under development but Crystal Systems have succeeded in producing 4 kg single crystal ingots of 15 cm x 15 cm cross-section from metallurgical grade feedstock.

Of course, all these approaches suffer from the disadvantage that the ingots have to be sawn into wafers. This is time-consuming and wastes up to 70% of the crystal, despite recent advances in sawing equipment. To avoid sawing altogether, many techniques are being developed to produce crystalline silicon in a continuous thin ribbon. This has obvious attractions as part of an automated plant. The main contenders in this field are:–

Self-Supported

Edge-defined film-fed growth (EFG) (Mobil-Tyco and IBM)	The ribbon is pulled upwards through a graphite die fed by capillary action from the melt, Mackintosh (1980) and co-workers
Dendritic web (Westinghouse)	A thin sheet of silicon is formed between two growing dendrites, which act as shaping guides, Duncan (1980) and co-workers.
Ribbon-to-ribbon (RTR) (Motorola)	A pre-formed ribbon of polycrystalline silicon is passed through a laser beam to enhance the grain size by local recrystallisation, Gurtler (1979) and co-workers.

Substrate-Supported

Silicon-on-ceramic (SOC) (Honeywell)	A slightly inclined substrate of aluminium silicate is passed over a narrow trough of molten silicon, Zook (1980).
Silicon-on-paper (RAD) (LEP, France)	A graphite-coated paper ribbon is pulled downwards through a slit in the bottom of a graphite crucible of molten silicon, Belouet (1980) and co-workers.

All these techniques are progressing slowly and the Japanese in particular are pressing on in this area. But the indications are that it will be some years before ribbons challenge sliced ingots in terms of throughput, reproducibility and cell efficiency.

To reduce cell fabrication costs, the present batch processes, such as the conventional junction diffusion and the vacuum deposition of contact metals, are being superseded by processes that can be automated as described by Van Overstraeten (1980). One of the most important developments in this field is junction formation by ion implantation. This is a low temperature process and affects only one side of the wafer, thus eliminating the back junction removal which is necessary with conventional diffusion. It can also be used to implant a back surface field (BSF), which increases open-circuit voltage and enhances the red response. In the USA, Spire Corporation have developed implanting equipment for processing 300 wafers/hour and have designs for plants with capacities of 10 and 100 MW/year according to the report of Kirkpatrick (1980) and co-workers. They have achieved efficiencies of 15% in single crystal cells and 10% in cast multicrystalline cells. Laser and pulsed electron annealing techniques are being perfected as an alternative to thermal annealing.

Diffusion from sprayed or spun-on dopants is another technique under development but reproducibility is a problem with these processes at present.

Screen-printed silver-based contacts were described by Michel (1980) and Baudry. They have already begun to replace the well-established vacuum-deposited Ti/Pd/Ag type in some factories. Although much cheaper, they are more susceptible to moisture, so an impermeable module package is essential. Lastly, sprayed-on and screen-printed single layer antireflective coatings are being developed with some success to replace the more expensive vacuum-deposited type of titanium oxide or tantalum oxide.

Several process sequences under current development are compatible with a module price of $1/Wp (1980 dollar).

5.6 Solar cells with concentrators

The use of concentrated sunlight is an important aspect of cost reduction, as it reduces the area of solar cells required and, up to a point, increases the cell efficiency. Moreover, the heat from the cell cooling system can be put to good use in some applications. Attempts to adapt silicon cells for concentrated sunlight have met with considerable success. The main feature of such cells is a very fine front contact grid to reduce internal resistance. Sandia Laboratories of Albuquerque in New Mexico recently claimed to have achieved an efficiency of 20% at 50 suns and 16% has been reached at 200 suns. This was described by Weaver (1980) and Nasby. The photographs in Figure 5.6.1 and Figure 5.6.2 show examples of concentrating solar cells at Sandia Lab., in 1979.

For higher concentrations, gallium arsenide cells are better, as they are more efficient and less temperature-dependent than silicon, although more expensive. A conductive GaAlAs window overcomes three of the problems encountered in the original GaAs solar cells, namely high surface recombination velocity, high sheet resistance and low shunt resistance. Efficiencies have reached 25% at 180 suns according to Sahai (1978) and co-workers, and 23% at 1000 suns were achieved by Vander Plas (1978) and co-workers.

Significantly higher efficiencies can theoretically be obtained through the use of multijunction or cascade structures having layers of materials with different energy gaps. For example, it is theoretically possible to achieve 40% at 1000 suns with a 2-junction cascade of GaAlAs and GaInAs but performances approaching this level have not yet been reported.

Various types of concentrator are being developed and field tested for photovoltaic systems. Among them are parabolic reflective troughs and

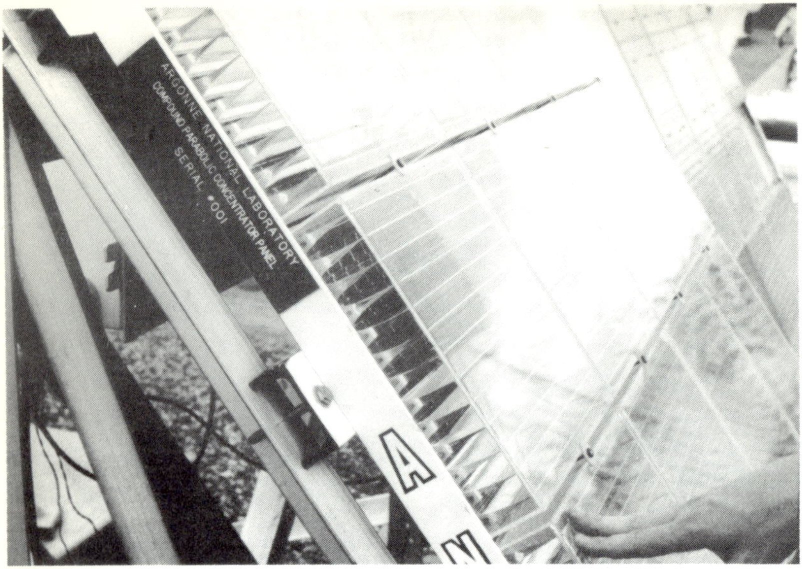

Figure 5.6.1: Solar cells with parabolic reflective troughs at Saudia Laboratories (1979).

Figure 5.6.2: Linear Fresnel lenses act as concentrators for photovoltaic cells at Saudia Laboratories in New Mexico (19797.

Figure 5.6.3: Tracking Concentrator.

linear Fresnel lenses for concentrations of up to 30 and point-focusing paraboloid and Fresnel lenses for higher concentrations. These devices, of course, are only effective under direct solar irradiation and except for very low concentration ratios, they need to track the sun. Such a device is shown in Figure 5.6.3 which is operated at Sandia Laboratories. The solar sensor is on the top of the back plate, the cells are in the focus of the parabolic mirror and the whole unit tracks around two axes. The higher the concentration, the more accurate the optical profile and tracking system have to be and the more sophisticated the cooling system. Thus the cost of the concentrator rises with concentration ratio and there is an optimum at which the total system cost will be a minimum. One European study has placed the optimum concentration ratio of about 30 but this will diminish as cell prices fall. Installed concentrator array prices have not dropped below $10/Wp.

It is interesting to note that the Dow Corning Corporation in the US has developed a 50 W concentrating solar electric generator X9-8501 which is marketed in developing countries. Currin (1978) described it during the Middle East Solar Technology Exhibition and Conference in Bahrain, which is in one of the high solar irradiance areas of the world.

5.7 **Thin-Film Cells**

Cost reduction to the point where photovoltaic generation can start to compete with central power stations will depend on a breakthrough in thin-film solar cell technology.

The main contender in this field is the cuprous sulphide-cadmium sulphide heterojunction cell. There are two versions – the frontwall, where the light enters the cuprous sulphide layer first, and the backwall, in which the cadmium sulhpide is illuminated. Although invented at about the same time as the silicon cell, this type has developed much more slowly, because of problems with low efficiency, instability and poor reproducibility. But it is potentially cheap because material costs are low and it is adaptable to mass production.

A group of research scientists, amongst them Hall (1978) and co-workers at the University of Delaware recently achieved an efficiency of over 10% in the Laboratory with a frontwall cell incorporating zinc in the CdS layer. In Europe, the University of Stuttgart has made 2000 units of 42 cm^2 frontwall cells averaging 7.3% without antireflective coating, which have shown no degradation after 560 days on load. Nukem are planning to put these Stuttgart cells into production. Two American firms, SES and Photon Power, are in pilot production with CdS cells but, to date (1981), their efficiencies are both under 3%. In the UK, Newcastle Polytechnic, The University of Durham and Thorn-EMI are working in this field.

The other important thin-film candidate is hydrogenated amorphous silicon (a-Si), which was pioneered by the University of Dundee. This material is an order of magnitude more absorbing than crystalline silicon, so only 1 or 2 microns are needed to make a solar cell.

RCA, who are developing Dundee's glow discharge process for depositing the film claim to have made rapid progress in 1980 having made hundreds of cells with efficiencies of over 5% which were described by Carlson (1978). However, the area of these cells is small (1 cm^2) and the efficiency goes down as the area is increased. The Japanese claim to have achieved 3.4% in 10 cm square cells. They have made prototype modules and plan to develop mass production techniques. Sanyo are already marketing pocket calculators powered by a-Si cells. The main obstacles to higher efficiency are high series resistance and the difficulties of controlling the fabrication of the doped layers.

In the UK, work on a-Si solar cells is being carried out at the Universities of Dundee and Sheffield, Heriot-Watt University, Edinburgh, the Royal Military College of Science, Shrivenham, the University of Manchester Institute of Science and Technology and the Plessey Co.'s Allen Clark Research Centre, Caswell. Sheffield are investigating an alternative technique, rf sputtering, for depositing the semiconductor.

Another potentially low cost approach is the metal-insulator-semiconductor or MIS cell, sometimes known as the Schottky barrier type. The metal can be either in the form of a substrate supporting a thin film of semiconductor or a thin, transparent layer on a self-supporting semiconductor.

Research on MIS cells is in the early stages. The control of the thickness and characteristics of the very thin oxide layer which is essential for adequate performance is posing problems. Many of the fabrication techniques are at present batch-type vacuum processes which are not amenable to large-scale production. Degradation mechanisms are still proving troublesome. The best reported thin-film MIS efficiency in 1981 is 4.8% from a 2 mm diameter cell consisting of 1000 A° of silver on a thin layer of GaAs deposited on a germanium-coated tungsten substrate as described by Yeh (1978) and co-workers. This cell has no antireflective coating. Most of the British work on amorphous silicon is centred on MIS structures. New work on materials for MIS devices which offer cost reduction has been reported by Singh (1980) and Leslie.

In the continuing search for a really cheap solar cell, many other types are being studied, mainly by University groups. For example, heterojunctions embodying II-VI compounds have the potential advantage of a direct near-optimum band gap and ease of fabrication by a variety of deposition methods. Among those being tried are combinations of absorber materials with a bandgap between 1 and 1.5 eV, such as p-type Si, GaAs, InP, CdTe or $CuInSe_2$ with a wide bandgap window layer of n-type CdS, ZnCdS, ZnSe, ZnO or indium tin oxide (ITO). Efficiencies of such structures are still under 3%, however, and none has yet emerged with any particular promise.

In summary, it appears that crystalline silicon will continue to dominate the solar cell field over the next decade at least and that silicon modules costing $1 to $2/Wp in 1980 terms may be with us before 1990.

5.8 **Future Applications**

Perhaps the most important application of photovoltaics in the near future is the provision of pumped water for human and animal consumption and for irrigation, particularly in the developing countries. 250 millions of the world's poorest people live in areas which experience extended dry seasons, during which crop production is impossible without irrigation. With proper land management and water resources, food production could be at least doubled. Apart from irrigation, life in many parts of the world could be transformed with pumped water. For instance, a 20 m well in the Sahara can provide 20 m^3 of water per day, enough to sustain some hundreds of people and as many head of cattle.

Water pumping, disinfection and desalination are very suitable applications for photovoltaics because, with an adequate reservoir, pumping can be restricted to the daylight hours and no electrical storage is required. Experience with the 60 or so photovoltaic water pumps already operating in Africa and other parts of the world has shown them to be simpler, more reliable and easier to maintain than other types, including solar thermodynamic and wind driven pumps. In Mali Vespieren (1980) reported that photovoltaic pumps are now serving 4500 people and saving 1000 litres of gas oil every month. Such failures as have occurred have been due to early module encapsulation faults, insect attack on exposed resin surfaces, cleaning problems, ''hot-spot'' failures, connection faults and lack of lightning protection. System designers have learned from this experience.

The photograph on Figure 5.8.1 shows a 616 Wp photovoltaic array powering a 500 W water pump in Corsica. The flow averages 16 m^3/day from a 25 m well.

About 80% of the population of the developing countries, some 2000 million people, live in rural areas without electricity. Nearer home, there are many isolated communities, for example in the Western Isles of Scotland and the Mediterranean, who either have no electricity or are supplied expensively by diesel generators. Solar cells offer an attractive solution to this problem. The provision of solar electricity on quite a modest scale might halt the population drift to the towns and encourage skilled people to move to the villages, with great benefit to the economy. For example, a 3 kWp array, extending over an area of 30 to 40 m^2, could provide basic lighting, refrigeration, educational TV and pumped water for 50 African houses. Wolfe (1981) of Lucas BP Energy Systems calculates that, even at present prices, photovoltaic systems are cheaper in the long run than alter-

Figure 5.8.1: Water pump in Corsica is powered by photovoltaic array visible in the background.

natives for average loads smaller than 200 to 2000 W, depending on local conditions. The break-even point will, of course, rise as solar cell prices fall. In areas with little technological skills available solar cells offer the advantage of low maintenance requirements, and this may accelerate their market penetration.

5.9 Grid Systems

Solar cells can also be considered as a contributory source of power in industrialised countries. This is particularly true for the US where there is a summer peak load for air-conditioning. The US Department of Energy predicts that, if and when system prices fall to \$1.6 to \$2.2/Wp, localised roof-top arrays, generating power where it is consumed and using the grid as a buffer store, will begin to be used on a significant scale for residences and so-called intermediate load centres such as hospitals, schools, shopping centres, offices and factories. Even in the UK, a 50 m^2 roof array would be sufficient to generate an average of 10 kWh/day, which is more than the average household uses. Each consumer would have his own power conditioning unit to enable the array to operate always at its maximum efficiency and convert the DC to AC of a waveform suitable for synchronous connection to the grid.

The photograph on Figure 5.9.1 shows a 5 kWp roof-top array, which was exhibited by Arco Solar at the IEEE Photovoltaic Specialists' Conference in San Diego in January 1980. It had a total area of 57.3 m^2 and a module efficiency of about 9%.

Figure 5.9.1: Arco Solar rooftop array in San Diego, 1980.

In the USA, the market for domestic roof-top arrays is being stimulated by a 40% tax rebate on the first $10,000 spent and by recent legislation which requires utility companies to pay the householder for excess electricity fed into their grid.

Table 5.9.1 shows the conventional and on-site photovoltaic energy prices predicted for 1986 by Clorfeine (1980) for residential and selected intermediate load centres in three American towns, assuming the $1.6/Wp system price goal is met. The conventional prices are based on an assumed annual escalation of 3% above inflation.

Table 5.9.1 Predicted Conventional and On-Site PV Energy Prices in 1986

Load	Location	Predicted Conventional Energy Price c/kWh	Photovoltaic Energy Price if DOE goal is met c/kWh
Residential	Phoenix Miami Boston	5.7 5.5 9.4	5.2 6.9 8.7
Selected Inter-mediate	Phoenix Miami Boston	6.4 7.0 8.0	5.5 7.3 9.2

The figures suggest that, if the DOE goal is met, on-site photovoltaics could compete over much of the country. Note that, although Boston has less sunshine than Miami, it is more favourable for solar electricity because of the higher price of conventionally generated electricity. This has some significance for the UK. Indeed, Wolfe (1981) has predicted that on-site solar power will become cheaper here than mains electricity in the 1990's.

The balance-of-system costs of central photovoltaic power stations are likely to be higher than in the case of on-site generation, as the land has to be paid for, the site made secure and transmission lines laid to the grid. Moreover, in the absence of storage batteries, cost calculations must take into account the back-up conventional generating capacity that must be provided if grid reliability is not to be adversely affected. When these things are taken into consideration, it works out that the module price must fall to between 15 and 40 c/Wp to compete with conventional coal, oil or nuclaer power stations. Before this can happen, there will have to be a breakthrough in thin-film solar cell technology, as has already been pointed out.

An extension of the central power station idea is the solar power satellites (SPS), a concept first proposed in 1968 and reported by Glaser (1973). He envisaged a system of power satellites in geostationary orbit, where an irradiance of nearly 1400 W/m^2 is continuously available, except for short periods of eclipse at the equinoxes, and the environment is relatively benign. In NASA's 5 GW Reference System, which has been used as a basis for assessment, each satellite has a 55 km^2 solar array with a microwave antenna to transmit the energy to one or more receiving stations on the ground, where it is converted to AC and distributed. All of the many studies that have been made so far agree on the technical feasibility of the concept, though many difficulties remain to be overcome and assumptions confirmed. Among the problems are that of finding sites for the 100 to 200 km^2 receiving stations and integrating a 5 GW station into the grid with the necessary back-up. The launcher would have to be much bigger than the present Space Shuttle. The strategic implications of placing a country's power source in space would have to be taken into consideration.

5.10 **Pilot Projects and Demonstrations**

Many pilot plants have been and are being set up to demonstrate the many future applications of photovoltaics, develop system design and test the performance and reliability of the hardware. The USA have had rural electrification projects running at Schuchuli, Arizona, since 1978 and at Tangaye, Upper Volta since 1979. They are now building a 350 kWp photovoltaic generator in Saudi Arabia to provide power to two villages, as part of the US-Saudi SOLERAS project. They also have an extensive programme for grid-connected residential and intermediate load applications, some using flat-plate arrays and some concentrating.

The World Bank, acting on behalf of the United Nations Development Programme, is funding a current project to assess small solar pumps for irrigation in developing countries. Field tests are being conducted in Mali, Sudan and the Philippines and are being complemented by design studies and laboratory tests on components and sub-systems. The ultimate aim is a robust, reliable, easily-maintained system capable of providing a flow of at least 1 litre per second against heads of up to 6 m at an overall cost of no more than 5 c/m^3. This corresponds roughly to a module price of \$2.80/Wp. The contractors for this project are Sir William Halcrow and Partners of London and Swindon.

The EEC Commission has recently launched its own photovoltaic pilot projects programme, under which 19 plants are to be built in various parts of Europe. Peak power ratings will range from 30 to 300 kWp and the plants will serve a variety of purposes, among them rural and island electrification, water pumping, disinfection and desalination, ice making for an agricultural cold store, TV and radio transmitters, a dairy farm, a vacation centre, power management and control at an airport and hydrogen production for a semiconductor factory. One of the plants will be an 80 kWp array feeding power to the grid at Marchwood Power Station, Southampton.

As examples, Figure 5.10.1 shows a model of the proposed 100 kWp scheme at the village of Vester Boegebjerg, Denmark, and Figure 5.10.2 shows a 300 kWp array for the vacation centre on Pellworm Island, Germany. All plants are due to be completed by mid-1983.

Figure 5.10.1: Model of 100 kWp rural electrification project in Vester Boegebjerg, Denmark *(Courtesy: EEC Commission).*

Figure 5.10.2: Model of 300 kWp array for a vacation centre on Pellworm Island, Germany *(Courtesy: EEC Commission).*

References

Adams, W. G. (1877) and R. E. Day: "The action of light on selenium." **Proc. Roy. Soc. London, Ser. A 25,** 1877, 113.

Anagnostou, E. (1978) and A. F. Forestieri: "Endurance testing of first generation (Block 1) commercial solar cell modules". **Proc. 13th IEEE Photovoltaic Specialists' Conference,** June 1978, 843-846.

Backus, C. E. (1980): "Principles of photovoltaic conversion in **Solar Energy Technology Handbook, Part A. Engineering Fundamentals.** ed. W. C. Dickinson and P. N. Cheremisinoff, Marcel Dekker 1980, 483-515.

Baikie, P. E. (1978): "Battery storage in solar cell systems", in **Storage in Solar Energy Systems.** Conference (C15) London, UK-ISES 1978, 68-75.

Becquerel, E. (1839): "On electric effects under the influence of solar radiation." **Compt. Rend. 9** 1839, 561.

Belouet, C. (1980), C. Berlin, J. Schneider and J. Paulin: "Progress in the growth and performance of RAD silicon sheets". **Proc. 3rd EC Photovoltaic Solar Energy Conference,** Cannes 1980, D. Reidl. 558-562.

Budyko, M. I. (1958): The heat balance of the earth's surface. English translation by N. Steponova, US Department of Commerce Weather Bureau, Washington DC, 1958.

Carlson, D. E. (1978): "The status of amorphous silicon solar cells". **Proc. 13th IEEE Photovoltaic Specialists' Conference,** June 1978, 294-301.

CEC (1981): **Recommendations for the Qualification Testing of Terrestrial Photovoltaic Modules,** Specification No. 201, to be published by the CEC Joint Research Centre, Ispra, Italy in 1981.

Chapin, D. M. (1954), C. S. Fuller and G. L. Pearson: "A new silicon p-n junction photocell for converting solar radiation into electrical power". **J. Appl. Phys. 25** 1954, 676-677.

Clorfeine, A. S. (1980): "economic feasibility of photovoltaic energy systems". **Proc. 14th IEEE Photovoltaic Specialists' Conference.** January 1980, 986-989.

Currin, C. G. (1978): "50 watt solar electric generator". in **SOLTECH '78, Conference Proceedings** of the Middle East Solar Technology Exhibition & Conference Bahrain, April 1978, UK-ISES, 57-60.

Darlington, A. D. (1978), P. T. Davies and J. L. J. Rosenfeld: "Estimation of size requirements for solar photovoltaic systems". in **Storage in Solar Energy Systems.** Conference (C15) London, UK-ISES 1978, 59-67.

Duncan, C. S. (1980), R. G. Seidensticker, J. P. McHugh, R. H. Hopkins, M. E. Heimlich, J. M. Driggers and F. E. Hill: "Development of processes for the production of low cost silicon dendritic web for solar cells". **Proc. 14th IEEE Photovoltaic Specialists' Conference,** January 1980, 25-30.

Freiesleben, W. (1980): "The solar material market – projections, needs and commitments". **Proc. 3rd EC Photovoltaic Solar Energy Conference,** Cannes 1980, D. Reidl, 166-170.

Glaser, P. E. (1973): "Space solar power". in **The Sun in the Service of Mankind.** UNESCO Conference, Paris 1973.

Grondall, L. O. (1933): "The copper-cuprous oxide rectifier and electric cell". **Rev. Mod. Phys. 5,** 1933, 141.

Gurtler, R. W. (1979), A. Baghdadi, R. N. Legge and R. J. Ellis: "Potential for improved silicon ribbon growth through thermal environment control." **Proc. 2nd EC Photovoltaic Solar Energy Conference,** Berlin 1979, D. Reidl, 145-152.

Hall, R. B. (1978), R. W. Birkmire, J. E. Phillips and R. D. Meakin: "Thin film polycrystalline (CdZn) S/CU_2S solar cells of 10% conversion efficiency". **Proc. 13th IEEE Photovoltaic Specialists' Conference,** June 1978, 1094-1096.

Hunt, L. P. (1979) and V. D. Dosaj: "Progress of the Dow Corning process for solar grade silicon". **Proc. 2nd EC Photovoltaic Solar Energy Conference,** Berlin 1979, D. Reidl, 98-105.

Jensen, J. (1981) and B. C. Tofield: "Advanced batteries for energy storage", Papers presented at the **International Conference on Energy Storage,** Brighton 1981, BHRA Fluid Engineering, Vol. 1, 205-216.

JPL (1981): **Block V Solar Cell Module Design and Test Specification for Residential Applications.** Prepared for the US Department of Energy by Jet Propulsion Laboratory, Pasadena, CA, January 1981.

Kachare, A. H. (1980), F. M. Uno, T. Miyahira and R. L. Lane: "Performance of silicon solar cells fabricated from multiple Czochralski ingots grown by using a single crucible". Proc. **14th IEEE Photovoltaic Specialists' Conference,** January 1980, 327-331.

Kirkpatrick, A. R. (1980), J. A. Minnucci and A. C. Greenwald: "Low-cost ion implantation and annealing technology for solar cells". **Proc. 14th IEEE Photovoltaic Specialists' Conference,** January 1980, 820-824.

Lange, B. (1930): "New photoelectric cell". **Zeit. Phys. 31,** 1930, 139.

Lindmayer, J. (1980): "Industrialization of photovoltaics". **Proc. 3rd EC Photovoltaic Solar Energy Conference,** Cannes 1980, D. Reidl, 178-185.

Lutwack, R. (1980): "A US view of silicon production processes". **Proc. 3rd EC Photovoltaic Solar Energy Conference,** Cannes 1980, D. Reidl, 220-227.

Mackintosh, B. (1980), J. P. Kalejs, C. T. Ho and F. V. Wald: "Multiple EFG silicon ribbon technology as the basis for manufacturing low cost terrestrial solar cells". **Proc. 3rd EC Photovoltaic Solar Energy Conference,** Cannes 1980, D. Reidl, 553-557.

Michel, J. (1980) and H. Baudry: "Reliable screen-printed contacts on silicon solar cells". **Proc. 3rd EC Photovoltaic Solar Energy Conference,** Cannes 1980, D. Reidl, 679-683.

Prince, M. B. (1955): "Silicon solar energy converters". **J. Appl. Phys. 26,** 1955, 534-540.

Rappaport, P. (1959): "The photovoltaic effect and its utilization". **RCA Rev. 20,** 1959, 373-397.

Reed, W. H. (1978), T. N. Meyer, M. G. Fey, F. J. Harvey and F. G. Arcella: Development of a process for high capacity arc heater production of silicon". **Proc. 13th IEEE Photovoltaic Specialists' Conference,** June 1978, 370-375.

Reynolds, D. C. (1954), G. Leies, L. L. Antes and R. E. Marburger: "Photovoltaic effect in cadmium sulphide." **Phys. Rev. 96,** 1954, 533-534.

Sahai, R. (1978), D. D. Edwall and J. S. Harris: "High efficiency AlGaAs/GaAs concentrator solar cells". **Proc. 13th IEEE Photovoltaic Specialists' Conference,** June 1978, 946-952.

Schmid, F. (1980), M. Basaran and C. P. Khattack: "Directional solidification of mg silicon by heat exchanger method (HEM) for photovoltaic applications". **Proc. 3rd EC Photovoltaic Solar Energy Conference,** Cannes 1980, R. Deidl, 252-256

Schottky, W. (1930): "Cuprous oxide photoelectric cells". **Zeit. Phys. 31,** 1930, 913.

Shumka, A. (1978) and K. H. Stern: "Some failure modes and analysis techniques for terrestrial solar cell modules". **Proc. 13th IEEE Photovoltaic Specialists' Conference,** June 1978, 824-834.

Singh, R. (1980) and J. D. Leslie: "Economic requirements for new materials for solar photovoltaic cells". **Solar Energy 24,** 1980, 589-592.

Van Overstraeten, R. J. (1980): "Advances in silicon solar cell processing." **Proc. 3rd EC Photovoltaic Solar Energy Conference,** Cannes 1980, R. Deidl, 257-262.

Vander Plas, H. A. (1978), L. W. James, R. L. Moon and N. J. Nelson: "Performance of AlGaAs/GaAs terrestrial concentrator solar cells". **Proc. 13th IEEE Photovoltaic Specialists' Conference,** June 1978, 934-940.

Vespieren, B. (1980): "The application of photovoltaics to water pumping and irrigation in Africa". **Proc. 3rd EC Photovoltaic Solar Energy Conference,** Cannes 1980, D. Reidl, 439-445.

Weaver, H. T. (1980) and R. D. Nasby: "High efficiency silicon solar cells". **Proc. 3rd EC Photovoltaic Solar Energy Conference,** Cannes 1980, R. Reidl, 1098.

Wolfe, P. R. (1981): "Photovoltaic systems for power generation". **Proc. Third International Conference on Future Energy Concepts,** January 1981, IEE Conf. Pub. No. 192, 165-168.

Wysocki, J. J. (1960) and P. Rappaport: "Effect of temperature on photovoltaic solar energy conversion". **J. Appl. Phys. 31,** 1960, 571-578.

Yeh, Y. C. M. (1978), F. P. Ernest and R. J. Stirn: "Progress towards high efficiency polycrystalline thin-film GaAs AMOS solar cells". **Proc. 13th IEEE Photovoltaic Specialists' Conference,** June 1978, 966-971.

Zook, J. D. (1980): "Evaluation of silicon-on-ceramic material for low cost solar cells". **Proc. 3rd EC Photovoltaic Solar Energy Conference,** Cannes 1980, D. Reidl, 569-573.

Chapter 6

BIOMASS

The biological conversion of solar energy is well understood by everyone on the simplest level since we have all personally experienced the effect of sunlight on plant growth. On a scientific level it is a complex process which has recently received more attention as the prospects of energy farming are being seriously considered.

Photosynthesis occurs in the chloroplasts of green plants and involves the conversion of solar energy into chemical energy in the form of carbohydrates (CH_2O). The basic chemical step is the combination of carbon dioxide (CO_2) and water (H_2O) by abstraction of oxygen (O_2). This can be represented in a simplified form as

$$CO_2 + H_2O + \text{solar energy} \rightarrow (CH_2O)_n + O_2)$$

The photosynthetic efficiency is dependent on the wavelength of radiation and can reach 9% for the visible region 500-700 nm in the green alga **Chlorella** according to Emerson (1943) and Lewis.

A review of bioconversion was given recently by Ward (1980) where the famous Calvin cycle tracing the pathway of carbon fixation is described.

In many developing countries biomass provides almost the only source of energy for the rural population and in the African continent for example, according to Howe (1980) and Gulick, it provides the greater proportion of all energy used. However, although the contribution of fuel wood to the global budget may be over 14% of the total consumption of energy, in general this is not planned.

A world overview of biomass was given during an ISES conference in London by Hall (1979) who surveyed amongst others the solar energy conversion processes, resources and products as well as the current status of R & D programs in the USA, Canada, Brazil, Europe, Sahel region, India, Philippines, Australia, New Zealand, China, South Korea and gave numerous references for further studies.

The largest and most spectacular rural energy programme is that in China described by El-Din (1980). It is based on small scale anaerobic digesters of which over 7 million have been built since 1972, as reported by Chen (1981). The largest biomass programmes directed at the production of commercial fuel are those in Brazil and the United States, the major product of which is ethanol produced by fermentation. Brazil leads the world in the gasohol programme. According to Gochnarg (1979) the Brazilians are on target to produce 11 billion litres of ethanol yearly by 1985. In the USA over 500 million gallons of ethanol will be produced in 1981 from maize. The International Molasses Report has included, over the last two years, IMR (1980/81), details of new power alcohol factories or programmes in Zambia, Zimbabwe, The Philippines, Nicaragua, Paraguay, Thailand, Indonesia and Kenya amongst others. Studies have been carried out in Australia, Stewart (1979) and co-workers, and in New Zealand, University of Auckland, (1979). Japan has a "green energy" programme and France has just announced an ambitious biomass based "carburol" programme to produce liquid fuels, Lloyd (1981).

The European Community has a programme, Chartier (1981), as does the United Kingdom where it is co-ordinated through the Energy Technology Support Unit, King (1979). The Canadian government supports the $30 million energy from the forest (ENFOR) programme, Marshall (1979), which it hopes will replace up to 8% of Canada's total primary energy demand. Recent estimates by White (1980) and co-workers, suggest that in the European Community biomass could produce over 5 x 10^6 tons oil equivalent (toe), or about 4% of the energy demand. According to estimates by King (1979) biomass available with reasonable ease could also produce about 4% of the UK energy demand, some 2.6 x 10^6 toe, although if all available land were cultivated this might be raised to around 18 x 10^6 toe. The recent estimate for the USA published by OTA (1980) suggests that energy from biomass could increase ten fold by the year 2000, and concludes that gasification of wood and methanol synthesis appears to be the best route for such energy utilisation.

Biomass in its strict sense could include all living matter. However, over the last decade it is the term which has become accepted to describe systems in which biological material is used not only as food, but also to produce a wide range of fuels, chemicals, fibre and building materials. At present the commercial emphasis is on the substitution of oil based liquid transport fuel by alcohols in particular. However, considerable research effort is directed at ways and means of achieving a much wider degree of substitution.

New methods of crop production and management are needed to give the required higher energy yields and new methods of conversion are needed in order to produce the required liquid fuel.

6.1 Biomass production as a solar process

Research on photosynthesis provides information as to the yields of biomass which might be expected at present, and in the longer term provides information for the plant breeder aiming at high yielding plants. It has also led to the initiation of projects aimed at producing cell free systems capable of producing hydrogen gas, for use as a fuel through photolysis of water, Bolton (1979) and Hall.

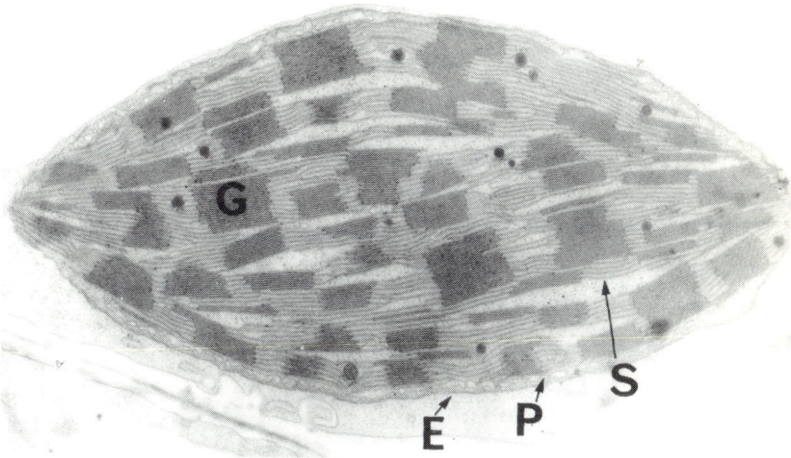

Figure 6.1.1: The chloroplast — the primary photoreceptor for biomass production through photosynthesis. This electron micrograph shows a section through a chloroplast from a mesophyll cell of a maize leaf. It is stained with osmium and lead, after glutaraldehyde fixation, to show the grana (G), outer envelope (E), peripheral reticulum (P) and non-membranous stroma (S). Magnification x 20,000. From Coombs (1976) and Greenwood.

To understand the limitations to biomass production it is necessary to consider both the details of the photosynthetic process and factors which affect the agricultural yield of biomass crops or trees grown on energy plantations. The photoreceptor for photosynthesis is the chloroplast. An electron micrograph showing a section through such an organelle is shown in Figure 6.1.1. Coombs (1976) and Greenwood have shown that these bodies, situated in the leaves of most higher plants and in a range of forms of algae contain the photoreceptive pigment chlorophyll. Light falling on the chlorophyll initiates an electron flow resulting in a potential gradient across the membrane. The flow of electrons is coupled to the production of an energy rich intermediate (ATP) and an organic reductant (NADPH). These products (NADPH and ATP) are used to power the fixation of atmospheric carbon dioxide and reduce it to the level of carbohydrate in dark reactions catalysed by enzymes located in the non-membrane region (stroma) of the chloroplasts. Further detail of the biophysical and biochemical aspects of these processes can be obtained, for example, from the volumes 5 and 6 of the Encyclopaedia of Plant Physiology edited by Trebst (1977) and Avron, and Gibbs (1979) and Latzko respectively.

6.2 Solar energy conversion efficiency

Not all the solar energy which reaches the surface of the earth can be used by green plants. In fact plants only use part of the radiation lying between 400 and 700 nm, as shown in Figure 1.1.1. This is termed photosynthetically active radiation (PAR) and has been established as about 43% of the total solar irradiance reaching the earth's surface, by Bassham (1977). Fixed CO_2, in the form of carbohydrate (CH_2O) has the energy content of 0.47 MJ/mol whereas the energy content of a mole quantum of red light is 0.176 MJ. Reduction of one molecule of CO_2 requires the flow of four electrons through two photosystems which operate in sequence, or a minimum of eight electrons. Due to limitations in the efficiency of the conversion process this number is probably closer to 10. Hence, 0.47 MJ are trapped as the result of the interception of 10 x 0.176 MJ and the efficiency will be the product of the 43% use as PAR multiplied by this fraction – or between 11 and 12%. However, the overall practical maximum efficiency of conversion of solar energy into organic compounds lies between 5 and 6%. This figure results from the fact that a significant proportion of the carbohydrate formed is subsequently consumed in dark respiration to yield the energy and reducing power necessary to form more complex and reduced plant compounds. The energy released by respiration is also used for the assimilation, reduction and incorporation of inorganic phosphorus, nitrogen and sulphur into organic compounds, as well as for the uptake of the micro-nutrients. In addition, in C3 plants, up to 30% of the carbon fixed by photosynthesis may be lost in photorespiration.

At the mean irradiance level which occurs in the UK, around 105 Wm^{-2}, and allowing for the cold winter months when plant growth is restricted, it might be expected crop yields would approach 100 tons of dry matter per hectare per annum. Actual yields of biomass are often less than 20% of this. The reason for this is that all plants are subjected to a variety of stresses which restrict yields. These stresses may be classed as environmental, nutritional or due to pests and diseases. Environmental constraints include availability of water and seasonal or daily variations in light and temperature. Nutritional limitations reflect in part the quality of soils, pH salinity, water retention capacity, etc., and in part the availability of fertilizers. The major classes of diseases and pests – bacteria, fungi, viruses, insects, etc. – are well known. The total effect of these stresses is to lower the actual yields of biomass to less than one percent of the available solar energy. The major requirement for efficient biomass production is that agricultural systems should be evolved to increase this value.

6.3 **Biomass for energy or food**

Most current agricultural practices are aimed at optimisation of good production. In many countries food production and/or availability is below that needed to support the existing population. As discussed by Brown (1980), this can lead to conflict of interest where current biomass fuel programmes which depend on potential food products as their raw materials.

A possible solution is to use non-food biomass resources such as wood, agricultural wastes, refuse and algae as raw materials. However, according to Smith (1981) conflicts of use may still arise since in many countries land and wood have restricted availability. Furthermore the factors limiting plant growth will still restrict the final yields of biomass. The measures needed to overcome these limitations may easily be listed. They include crop breeding for high yielding varieties and disease resistance; better water management and irrigation; use of fertilizers; mechanised tillage, harvesting and transport; improved storage systems, and so on. However, all these improvements in yield depend on an increased input of energy and cash into the system. Hence, yields per unit of land improve as yields per unit of energy input decrease.

At present varying levels can be recognised. According to Coombs (1980) these range from no inputs for a natural eco-system, harvested at intervals and allowed to regenerate naturally, through the labour intensive subsistence farm system typical of many tropical countries, with no inputs from sources external to the farm, to the highly mechanised, irrigated, high energy input systems as in the mid-west USA.

An unmanaged eco-system may result in a high yield where the natural vegetation is forest so long as care is taken to harvest it rather than destroy it by large scale de-afforestation. However, in many countries harvesting, or in worse cases destruction of the forest with no use of wood, is followed by subsistence agriculture. This is characterised by an absence of cash inputs, low crop yields and shifting cultivation often with destrcution of the soil, so that yields are static or falling with time. King (1978) and Chandler considered it possible that if well managed and integrated with tree planting such systems could provide a local population with both food and fuel.

6.4 **Energy and Agricultural Residues**

Criteria for selection of potential energy crops relate in part to the agricultural aspects just discussed, in part to the price at which they may be obtained and in part to the ease in which they may be upgraded to more valuable products. The many other potential biomass feedstocks currently being used on a restricted scale or researched, are shown in Table 6.4.1. These may be classified as land crops, aquatic crops, or waste residues and by-products. The required characteristics of high yield combined with low energy input are typified by forests and tree plantations.

Table 6.4.1 Sources of biomass for conversion to fuels

WASTES	**LAND CROPS**	**AQUATIC PLANTS**
Manures	**Ligno-cellulose**	**Water Weed**
Slurry	**Trees**	
Domestic rubbish		Water hyacinth
Food wastes	Eucalyptus	Water reeds/rushes
Sewage	Willow	
	Poplar	**Algae**
Residues	Alder	
	Firs, pines	Uni-cellular
Wood residues	Casuarina	Multi-cellular
Cane tops	Luceana	
Straw		
Husks	**Starch Crops**	
Citrus peel		
Bagasse	Maize	
Molasses	Cassava	
	Sorghum, Sago	
	Sugar cane	
	Cane	
	Beet	
	Hydrocarbon plants	
	Euphorbia	
	Croton	
	Hevea	
	Guayale	
	Oils	
	Babussu palm	
	Soyabean	
	Peanut	
	Sunflower	

As far as land plants are concerned trees probably represent the ultimate biomass crop. However, for sustained yields significant energy inputs may be needed. In general, in the developed countries, research on the use of trees has been carried out on short rotation coppiced plantations, Siren (1979), Steinbeck (1981). Szego (1973) and Kemp have been advocating energy plantation and their idea has been developed into an integrated bioconversion system which is called the Photosynthesis Energy Factory. This was described by Fraser (1979) and co-workers as consisting of a dry land energy plantation, a wood fired power plant and a biological waste water treatment system, like an algae pond. The products are electricity, synthetic natural gas and reclaimed waste water. Traditionally, large scale plantations of eucalyptus have been used in Brazil for charcoal to be supplied to the iron industry. The National Academy of Sciences (1977 and 1980) has published details of a wide range of under-exploited trees suitable for use in the production of rural firewood, and drawn attention to the value of the nitrogen fixing organism Leucaena. The availability of wood and agricultural residues on an annual basis exceeds the total oil demand, but it is unlikely that enough will ever be available in many regions, except in countries with a high per capita availability of wood such as Canada and the USSR as shown in Table 6.4.2. This table also shows the dramatic effect of unplanned use of fuel wood as has occurred in Asia, without adequate reafforestation.

Table 6.4.2 Wood availability by Country

	Closed Forest		Growing Stock of Closed Forests	
	(10^6ha)	Total ($10^6.m^3$)	$m^3.ha^{-1}$	m^3 per head
USA	220	20,200	92.1	95.3
Canada	250	17,800	71.2	809.1
Europe	133	14,900	103.5	29.1
USSR	770	81,800	106.2	324.6
Asia and Oceania	448	38,700	72.9	18.7
S. America	631	92,000	145.8	362.2
Africa	188	35,200	187.2	90.03

At present liquid fuels produced on a commercial scale from biomass come from sugar or starch crops. Coombs (1980) reports that sugar is the most important of these crops, Figure 6.4.1. Interest in hydrocarbon and oil crops stems from the fact that these products are more suitable for use in diesel engines, Hall (1980). Total availability of the major classes of carbohydrate and oil crops is shown in Table 6.4.3 where it is compared with net primary production of organic matter and wood production.

Figure 6.4.1: Sugar cane harvest.

Typical yields in developing and developed regions are given in Table 6.4.4. The figures for total global production may be compared with the current oil consumption of around 3 x 10^9 tons per year as reported by Moore (1979).

Table 6.4.3 Total production (10^8t)

Net primary production (organic matter)	2000
Forest production (dry matter)	900
Cereals (as harvested)	15
Cereals (as starch)	10
Root crops	5.7
Root crops (as starch)	2.2
Sugar crops	10
Sugar crops (as sugar)	0.9
Vegetable oils	0.5

Table 6.4.4 Total production (10^6t) and yields ($t.ha^{-1}$) of the major sugar and starch crop.

These figures are on a crop basis excluding the considerable amounts of crop wastes, straw, etc. The cereal figures exclude such crops grown for forage silage.

a) **Sugar**

	Cane		Beet	
	Total	Yield	Total	Yield
World	737	56	290	32
Developed countries	70	79	263	32
Developing countries	667	54	26	30

b) **Starch**

	Cereals		Maize		Rice		Wheat	
	Total	Yield	Total	Yield	Total	Yield	Total	Yield
World	1459	1.9	349	3.0	366	2.6	386	1.7
Developed countries	765	2.5	236	4.7	26	5.4	261	1.9
Developing countries	687	1.6	113	1.7	340	2.5	125	1.3

c) **Starch - root and tubers**

	Roots & Tubers		Potatoes		Sweet Potatoes		Cassava	
	Total	Yield	Total	Yield	Total	Yield	Total	Yield
World	570	11.0	272	15.0	138	9.6	110	8.8
Developed countries	225	15.7	222	15.7	2	14.7	–	–
Developing countries	345	9.2	70	10.4	136	9.5	110	8.8

Most aquatic crops such as reeds and other marsh plants discussed by Kresovich (1981) and co-workers, or giant kelp described by North (1980) are in the early stages of investigation although considerable experience exists in large scale algal production in open ponds associated with waste water treatment according to Shelef (1977) and co-workers.

Estimates as to the availability of agricultural residues have been made in many countries or regions and in some cases utilisation and conversion programmes initiated; these include Ontario (1980), the United Kingdom, King (1979), Denmark, Rexen (1976), and many studies in the United States such as for example Klass (1978).

6.5 **Conversion processes**

The major processes which are currently available or being researched for the conversion of biomass into more readily usable energy forms are shown in Figure 6.5.1. Such conversion is necessary for two reasons, first to upgrade the raw materials to products of higher energy and physical density and second to produce a product in a form in which it is compatible with technology which has largely been developed on the basis of the use of hydrocarbons as raw materials.

The conversion processes can be classified on the basis of the end products in terms of heat, process steam, electricity, gas of various forms, oil, char, alcohols and other chemicals, etc. It is also possible to classify them as thermal/chemical or as biological. The main differences between these are the characteristics of the raw materials suitable for such use. The major properties to be considered are the moisture content and concentration of low molecular weight fermentable solids. Crop residues such as bagasse can be burnt at 50% moisture content, but in general thermal conversion processes require raw materials with a lower (10 to 15%) water content. In contrast the biological processes use dilute feedstocks at concentrations of less than 20%.

Because of the overwhelming requirement to find substitute liquid fuels, most of the current work is directed towards alcohol production. A large number of reports concerning alcohol production have been published. These range from discussions of small scale ethanol production, SERI (1980), to regional (e.g. A Manual on Ethanol and Gasohol Production in Louisiana, LDNR (1980), national, Nathan (1978); University of Sydney, (1978); Stewart (1979) and co-workers, and international, World Bank (1980). Additional information can be obtained from publications such as the Bio-Energy directory, BEC (1980); report No. 5 of the Watt Committee (1979); UK-ISES (1979) and the proceedings of the recent Energy from Biomass conference sponsored by the Commission of the European Communities, Palz (1981) and co-workers. From these the key problems can be established as a reduction in both raw material costs and in the energy required to produce the ethanol in a form in which it can be used in car engines. These factors are considered in more detail in section 6.7.

Improvements in ethanol production technology range from studies on new systems of continuous fermentation, the use of new organisms through studies on immobilized cells alternative methods of alcohol separation, to direct fermentation on complex substrates.

The overall processes used in the major power alcohol programmes are summarised in Figure 6.5.2, and typified by the direct use of cane juice as

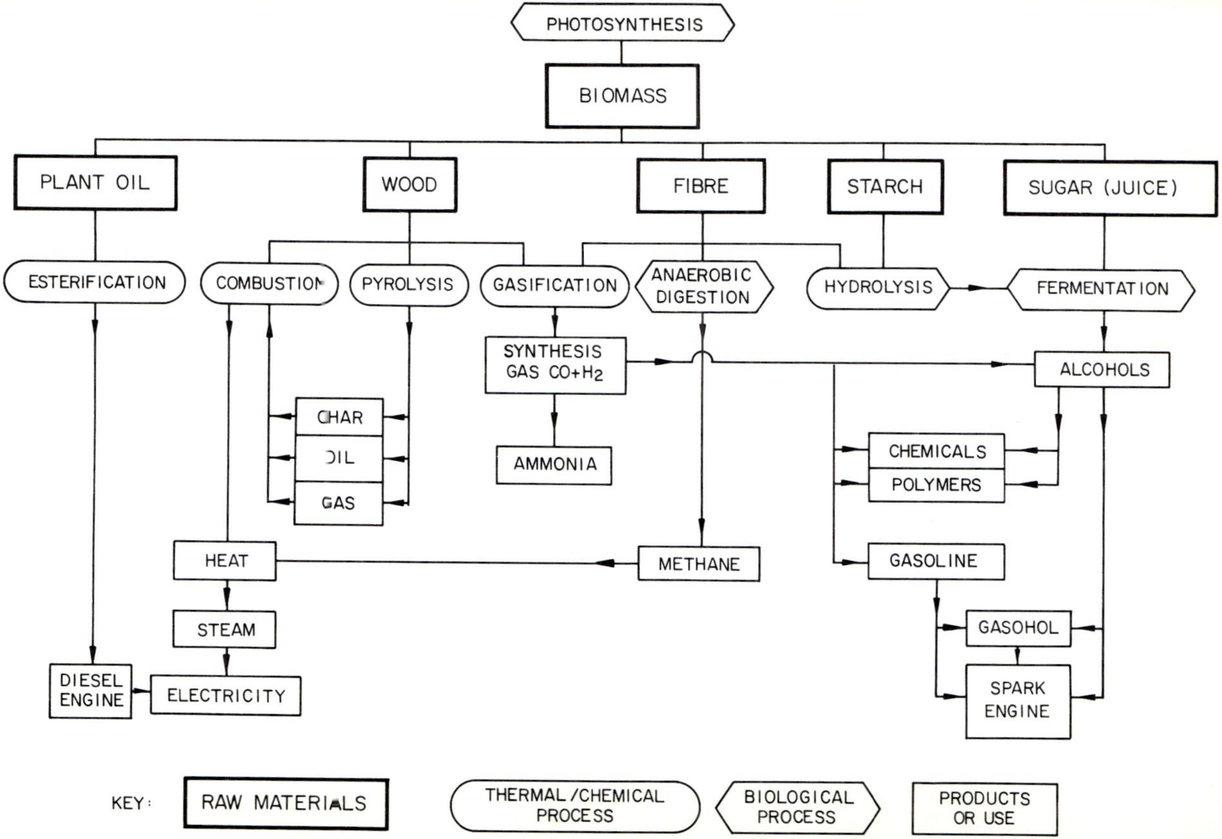

Figure 6.5.1: Processes used in the conversion of biomass into more readily usable energy forms.

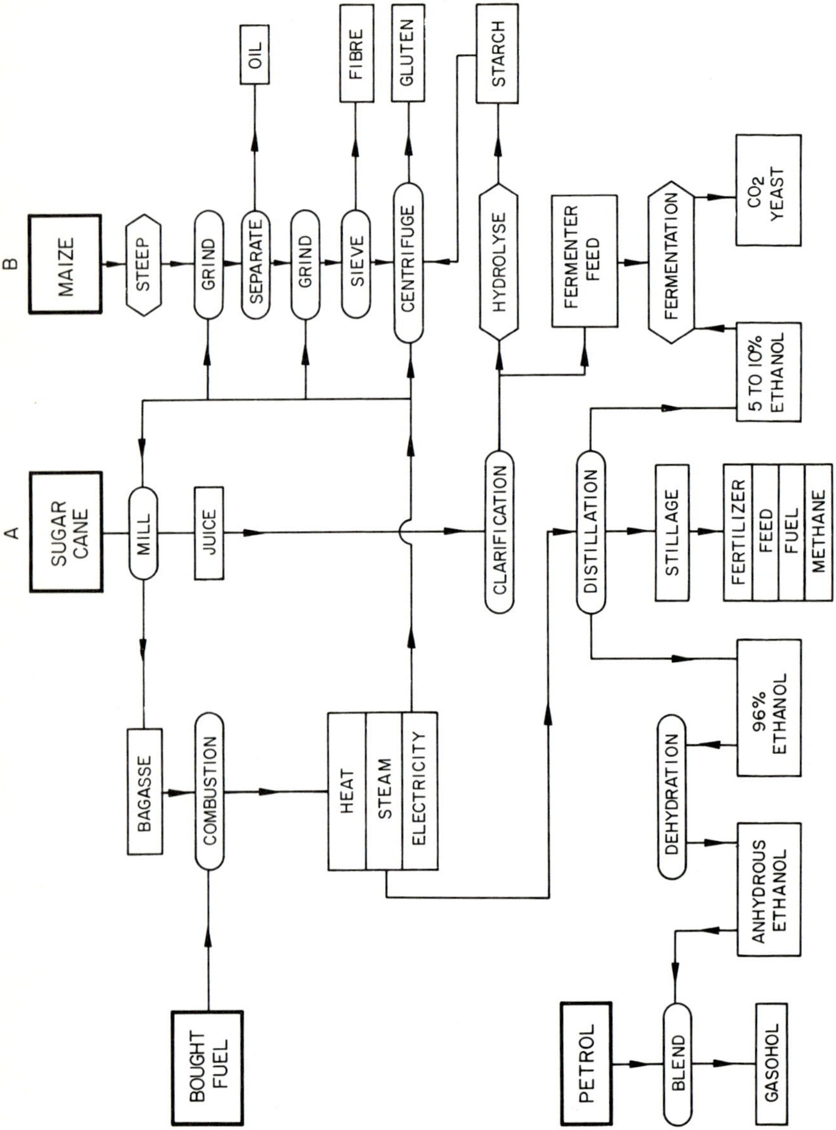
A
B
SUGAR CANE
MAIZE
STEEP
GRIND
SEPARATE
OIL
GRIND
SIEVE
FIBRE
CENTRIFUGE
GLUTEN
STARCH
HYDROLYSE
FERMENTER FEED
FERMENTATION
CO_2 YEAST
5 TO 10% ETHANOL
MILL
JUICE
CLARIFICATION
DISTILLATION
STILLAGE
FERTILIZER
FEED
FUEL
METHANE
BAGASSE
COMBUSTION
HEAT
STEAM
ELECTRICITY
96% ETHANOL
DEHYDRATION
ANHYDROUS ETHANOL
BOUGHT FUEL
PETROL
BLEND
GASOHOL

in Brazil, or fermentation of glucose syrup from maize as in the USA. In the case of cane-based systems all the energy for both extraction of the juice and separation of the alcohol by distillation may be obtained by combustion of the fibrous residues (bagasse). In contrast the maize based system requires an external source of fuel. This can be coal, natural gas or oil. However, since the total amount of energy expended to produce ethanol from maize may be close to its actual energy content, this is not a particularly viable proposition if the primary objective is oil substitution unless surplus grain and coal can be used. The third type of power alcohol plant is that which uses molasses, a by-product of sugar production, as raw material. Some factories using cassava have also been built.

The most abundant raw materials which could be used for alcohol production, and would get round problems of raw material costs and competition with food use, are wood wastes and agricultural residues. However, the problem at present is that no economic process for converting ligno-cellulose into a fermentable feedstock has yet been perfected. In its natural state cellulose in the plant cell wall exists as an insoluble complex with lignin which resists attack by both acid and enzymes. The plant cell wall also contains other substances (hemi-celluloses) which on hydrolysis release five carbon sugars which cannot be used by yeast. Hence, for ligno-cellulose to be used as a substrate for conventional yeast fermentation it must be milled and chemically or physically treated, by processes such as steam explosion, to disrupt the structure. Lignin may be removed by solvents or reagents such as sodium hydroxide, sodium hypochlorite, or peracetic acid. Alkalies also cause the structure to swell and fibres to separate. Mandels (1974) and co-workers have studied the hydrolysis of lignin using dilute sulphuric acid or enzymes from fungi. Much of the research in this area is carried out in the USA where it is co-ordinated, through the Solar Energy Research Institute who list major areas of work in a newsletter, SERI (1979 A). In particular, Ledisch (1979) and co-workers have studied complete systems in which in addition to alcohol production from glucose released from cellulose, the hydrolysis products of hemi-cellulose are used to produce for example butanediol, whereas the residual lignin is used to fuel the process. Emert (1980) and Katzen reported another approach in which the hydrolysis and fermentation is carried out in the same vessel, this overcomes the end product inhibition of enzyme activity caused by simple sugars. Developments in enzymes for cellulose hydrolysis have been reviewed by Ghose (1977). In spite of this considerable research effort economic processes of alcohol production from ligno-cellulose have yet to be developed. This is one of the major areas where a breakthrough could have a real impact on biomass energy programmes.

At present, the most efficient way of converting ligno-cellulose to a higher grade fuel (methane) is anaerobic digestion. A useful introduction to small scale digesters has been published by the National Academy of Science and a more detailed summary of the present status of research by Stafford (1980) and co-workers. As the required product just bubbles off, separation is easy, although for some purposes scrubbing to remove impurities may be required. Bacteria which break down cellulose work better at higher temperature. This may be a problem, particularly in temperate climates for example, in the use of manure in winter when the temperature of the incoming raw material is low. If a source of waste heat is not available a significant proportion of the gas produced may have to be burnt in order to heat this feed stream. In the UK commercial development in this area is being co-ordinated by the British Anaerobic Biomass Association Ltd. Keable (1979) and Dodson have studied different strategies to be used in gas production from farm waste. They proposed a modular system for bio-gas production based on a below ground anaerobic digester with sectional caps which can be removed sequentially with only a partial loss to production.

An alternative approach to the production of fuel alcohol is the formation of methanol derived from synthesis gas produced in a biomass gasifier. Synthesis gas is also the starting point for the production of a number of chemicals including ammonia. It may also be used to form a synthetic gasoline either via the traditional Fisher-Tropsch synthesis or by the new Mobil process, Penick (1978) and co-workers. The most comprehensive summary concerning the status of gasification of biomass is that of the Solar Energy Research Institute, SERI (1979 B). It is probable that a wood to methanol process will be perfected before a wood to ethanol fermentation process. This view is supported by the emphasis being put on gasification within the EEC, Palz (1981) and co-workers, and the UK, ETSU (1980).

A fluidised bed combustion process described by Stallings (1979) recovers low pressure steam from the combustion of wood chips.

6.6 **Liquid transport fuel**

The major objective of many biomass energy programmes is to provide fuel for use in an internal combustion engine. Both methanol and ethanol have high octane ratings and are suitable for use as fuels in spark ignition engines – although special designs may be needed. Ethanol in particular has been developed in this context since it may be used successfully in a spark ignition engine, either blended or as a fuel in its own right as described by Bernhardt (1981) as well as by Wagner (1979) and co-workers.

Alcohols may also be used as octane boosters, as in Petrol as in the US gasohol or German M15 – the first containing 10% ethanol and the second 15% methanol. Higher levels of alcohols may be added, as in Brazil or Zimbabwe for example, to achieve a greater degree of oil substitution at a national level. However, as the amount of alcohol is increased above that needed to boost the octane the mileage achieved per volume of fuel decreases due to the lower calorific value of the alcohols.

Usually alcohols cannot be used in an unmodified diesel engine. These engines can be modified by addition of a spark plug, or the fuel modified by blending with diesel distillate or a suitable ignition promoter. An alternative approach is to use plant oils, preferably modified by partial cracking or esterification. Pischinger (1981) and co-workers reported recently the results of running vehicles on vegetable oils and derived esters, from plants such as sunflower, rape, peanuts, soya, maize and palm. Considerable research in this area continues, although the cost of many such oils and the low yields will probably preclude widespread use.

Significant substitution of oil products in transport cannot occur using fuels derived from sugar and starch crops or from current production of plant seed and palm oils. However, in general, the actual amount of liquid transport fuel used in many developing countries is low as compared to that in the developed countries. For example, the total use in Africa, on a continental basis was around 60 million tons in 1978 as compared with 974 million tons in North America, according to Moore (1979). Hence, a significant proportional substitution can be made in the developing countries without a drastic requirement for increased land use.

6.7 **Energy balances and economics**

Most of the conversion processes have high energy yields in terms of the proportion of energy in the biomass raw materials which ends up in the product. For instance the theoretical yields in both anaerobic digestion and ethanol fermentation are over 90%. The problem as far as fermentation is concerned lies in concentrating the dilute 5 to 10% ethanol solution to the near anhydrous form needed for blending with gasoline. Considerable energy is expended in this distillation. The second factor relates to the size of the reaction vessel required for mass production, which in turn reflects the volumetric productivity of the process. Although volumetric productivities are less than 1 gram of ethanol per litre of reactor volume in conventional batch processes this may be increased by an order of magnitude by using techniques of cell concentration and recycle as described by Coombs (1981). In anaerobic digesters volumetric productivities are low because of the complex nature of the substrates used in many systems. 50% of the raw material may be digested with retention times of over 15 to 20 hours giving yields of some 0.3 m^3 of gas per kg of dry solids equivalent. In the case of methanol on the basis of experience with production from fossil fuels yields would probably approach 40% on a weight basis.

Multiplication of yields by energy content of the product will give an indication of the theoretical energy production for a given process. Unfortunately, this is not a measure of the actual energy gain, which will be much lower due to the considerable process energy put into the system, the energy used in transporting the raw materials to the factory, and the energy used in growing it in the first place. The gain may be considered in terms of the net energy ratio which considers the energy in the useful products to the total energy expended in their production. Values of over one indicate a net energy gain. According to Coombs (1980) such values are seen in ethanol fermentations where process energy is derived from crop residues such as bagasse, or other biomass fuels such as wood. The production of oil from plants shows a higher net energy ratio than the production of fuels by fermentation due to the lower inputs of process energy. However, with the possible exception of palm oil, the yields in terms of net energy harvested per unit load area are lower.

Many reviews including those by Nathan (1978); LDNR (1980); Stewart (1979) and co-workers, show that for alcohol production material costs represent between 60 and 85% of the total cost. An alternative way of looking at the same figures is to consider at what price the raw materials would have to be produced in order to make alcohol production competitive with oil purchase. The World Bank (1980) has concluded that sugar cane based production is likely to be economic, at the early 1980 oil price of $31 a barrel, providing the cost of cane is less than $14 a ton. Ethanol from molasses is economic where it can be produced for less than $60 per ton. However, cassava and corn based systems are less attractive since the fuel has to be purchased. To compensate for this, low price raw material is needed – cassava at less than $13 a ton and maize at less than $1 per bushel. In early 1981 in the US molasses were over $100 per ton, maize was over $3 a bushel and in many countries cane costs more than $30 a ton to produce. Production of alcohol from these sources continues, on the basis of the incentives found in the introduction of self sufficiency and savings on foreign exchange as a result of oil-substitution.

Fermentation alcohol produced from potential food materials is not the best, nor the cheapest biomass option. However, it is available now. It would appear safe to assume that in due course it will be supplanted by other products, probably derived from wood, as continuing biomass research results in innovations which will lower the product cost and increase the net energy gain.

References

Bassham, J. A. (1977): "Increasing crop production through more controlled photosynthesis" **Science, 197,** 1977, 630-637.

BEC (1980): **Bio-Energy Directory.** 2nd ed. Bio-Energy Council, Washington D.C. 1980.

Bernhardt, W. (1981): "Future automotive fuels" in **Energy from Biomass.** ed. W. Palz, D.O. Hall and P. Chartier, Applied Science Publishers, 1981.

Bolton, J. R. (1979) and D. O. Hall: "Photochemical conversion and storage of solar energy". **Annual Review of Energy, 4,** 1979, 353-401.

Brown, L. R. (1980): **Food or Fuel: New Competition for the World's Cropland.** Worldwatch Paper 35, Worldwatch Institute, Washington D.C., 1980.

Chartier, P. (1981): "Prospects for energy from biomass in the European community" in **Energy from Biomass,** ed. W. Plaz, D. O. Hall and P. Chartier, Applied Science Publishers, 1981.

Chen, R. C. (1981): "Biogas production in China". **Biomass 1** 1981 in press.

Coombs, J. (1976) and A. D. Greenwood: "Compartmentation of the photosynthetic apparatus" in **The Intact Chloroplast,** ed. J. Barber, Elsevier, 1976, 1-52.

Coombs, J. (1980): "Renewable sources of energy (carbohydrates)". **Outlook on Agriculture, 10,** 1980, 235-245.

Coombs, J. (1981): "Biogas and power alcohol". **Chemistry and Industry,** 4 April, 1981, 223-229.

El-Din, M. N. (1980): "Recycling of organic wastes in agriculture", in **Bioresources for Development,** ed. A. King and H. Cleveland, Pergamon Press, 1980, 184-203.

Emerson, R. (1943) and C. Lewis: "The dependence of the quantum yield of chlorella photosynthesis on wavelengths of light". **Am. J. Botany, 30,**1943, 165.

Emert, G. H. (1980) and R. Katzen: "Gulf's cellulose to ethanol process". **Chemtech,** October 1980, 610-614.

ETSU (1980): **Energy a Register of Research Development and Demonstration in the U.K. Part 3, Renewable Energy.** Energy Technology Support Unit, Harwell, HMSO 1980.

Fraser, M. D. (1979), J. F. Henry, L. C. Borghi and N. J. Barbera: "Analysis of the photosynthesis energy factory as an integrated bioconversion system", in **Sun II,** Proc. ISES Silver Jubilee Congress in Atlanta 1979, Pergamon Press, Vol. 1, 69-72.

Ghose, T. K. (1977): "Cellulose biosynthesis and hydrolysis of cellulosic substrates". **Advances in Biochemical Engineering, 6,** 1977, 39-76.

Gibbs, M. (1979) and E. Latzko, ed.: **Photosynthesis II. Carbon Metabolism and Related Processes.** Encyclopaedia of Plant Physiology. New Series. Vol. 6, Springer Verlag 1979.

Gochnarg, I. (1979): "The Brazilian national programme", in **Biomass for Energy,** Conference (C20) London, UK-ISES, 1979, 30-50.

Hall, D. O. (1979): "World biomass: an overview", in **Biomass for Energy,** Conference (C20) London, UK-ISES, 1979, 1-14.

Hall, D. O. (1980): "Renewable resources (hydrocarbons)". **Outlook on Agriculture, 10,** 1980, 246-261.

Howe, J. W. (1980) and F. A. Gulick, ed.: **Fuel Wood and Other Renewable Energies in Africa.** ODC Washington D.C., 1980.

IMR (1980/81): **International Molasses Report.** F. O. Licht, Germany. Vol. 17, 1980 and Vol. 18, 1981.

Keable, J. (1979) and C. Dodson: "A modular system for bio-gas production using farm waste", in **Sun II,** Proc. ISES Silver Jubilee Congress in Atlanta, 1979, Pergamon Press, Vol. 1, 83-87.

King, G. H. (1979): "UK Department of Energy, solar biological programme – biofuels", in **Biomass for Energy,** Conference (C20) London, UK-ISES 1979, 15-24.

King, K. F. S. (1978) and M. T. Chandler: **The Waste Lands.** International Council for Research in Agroforestry, Nairobi, Kenya, 1978.

Klass, D. C. (1978): "Energy from biomass and wastes", in Symposium Papers, Institute of Gas Technology, Chicago, August 1978, 1-28.

Kresovich, S. (1981), C. K. Wagner, D. A. Scantland and W. T. Lawhon: "Emergent aquatic plants, biological and economic perspectives". **Biomass 1,** 1981, in press.

LDNR (1980): **A Manual on Ethanol and Gasohol Production in Louisiana.** Louisiana Dept. of Natural Resources, P.O. 44396 Baton Rouge, Louisiana, 1980.

Ledisch, M. R. (1979), M. C. Flickenger and G. T. Tsao: "Fuels and chemicals from biomass". **Energy 4,** 1979, 263-275.

Lloyd, A. (1981): "France turns to alcohol for fuel". **New Scientist, 89,** 1981, 197.

Mandels, M. (1974), L. Hontz and J. Nystrom: "Enzymatic hydrolysis of waste cellulose". **Biotechnology and Bioengineering, 16,** 1974, 1471-1493.

Marshall, J. E. (1979): "Canadian biomass perspectives – a new interest in an old fuel", in **Biomass for Energy.** Conference (C20) London, UK-ISES, 1979, 51-59.

Moore, R. (1979): **Oil and Gas International Year Book.** ed. R. Moore, Financial Times, Longman, 1979.

National Academy of Sciences (1977): **Leucaena – Promising Forage and Tree Crop for the Tropics,** Washington D.C. 1977.

National Academy of Sciences (1980): **Firewood Crops.** Washington D.C. 1980.

Nathan, R. A. (1978): **Fuel from Sugar Crops.** D.O.E. Critical Review Series. Technical Information Center, U.S. Department of Energy, Springfield, Va. 1978.

North, W. J. (1980): "Biomass from marine macroscopic plants". **Solar Energy 25,** 1980, 387-395.

Ontario (1980): **Energy from Wastes, a Programme for Ontario.** Ontario Government Bookstore, 880 Bay Street, Ontario, Canada, March 1980.

OTA (1980): **Energy from Biological Processes.** Office of Technical Assessment, Congress of the United States, Washington D.C. 1980.

Palz, W. (1981), D. O. Hall and P. Chartier: **Energy from Biomass.** Applied Science Publishers, 1981.

Penick, J. E. (1978), S. J. Meisel, W. Lee and A. J. Silvestri: "Mobil process for conversion of coal and natural gas to gasoline", in **Alcohol Fuels.** Institution of Chemical Engineers, NSW Group, Sydney, Australia, 1978, 5-12.

Pischinger, G. (1981), R. Siekmann and F. Clymans: "Diesel oil substitution by vegetable oil – fuel requirements and vehicle experiences". I.E.A. Conference on New Energy Conservation Technologies. Berlin 6-10 April, 1981.

Rexen, F. (1976): "Straw as an industrial raw material", in **Solar Energy in Agriculture.** Conference (C9) University of Reading, 1976, UK-ISES, 39-43.

SERI (1979 A): **Biomass Refining Newsletter.** Solar Energy Research Institute, Golden, Colorado, 1979.

SERI (1979 B): **A Survey of Biomass Gasification.** Solar Energy Research Institute, Golden Colorado, 1979.

SERI (1980): **Fuel from Farms.** SERI/5P-451-519. US Government Printing Office, Washington D.C. 1980.

Shelef, G. (1977), R. Moraine, T. Berner, A. Levi and G. Oron: "Solar energy conversion via algal wastewater treatment", in **Photosynthesis 77,** ed. D. O. Hall, J. Coombs and T. W. Goodwin. The Biochemical Society, London, 1977, 657-675.

Siren, G. (1979): "Woodfuel experiments in Sweden", in **Biomass for Energy.** Conference (C20) London, UK-ISES 1979, 26-29.

Smith, N. (1981): **Wood, an Ancient Fuel with a New Future.** Worldwatch Paper 42, Worldwatch Institute, 1176 Massachusetts Ave., Washington D.C. 1981.

Stafford, D. A. (1980), B. I. Wheatley and D. E. Hughes: **Anaerobic Digestion.** Applied Science Publishers, 1980.

Stallings, J. W. (1979): "Economics of energy recovery from biomass using fluidized-bed combustion", in **Sun II,** Proc. ISES Silver Jubilee Congress in Atlanta, 1979, Pergamon Press, Vol. 1, 73-77.

Steinbeck, K. (1981): "Short rotation forestry as a biomass source", in **Energy from Biomass.** ed. W. Palz, D. O. Hall and P. Chartier. Applied Science Publishers, 1981.

Stewart, G. A. (1979), H. A. Nix, G. Gartside, W. H. M. Rawlins, R. M. Gifford and J. R. Siemon: **The Potential for Liquid Fuels from Agriculture and Forestry in Australia.** CSIRO Australia 1979.

Szego, G.C. (1973) and C. C. Kemp: "Energy forest and fuel plantation". **Chemtech,** May 1973, 275.

Trebst, A. (1977) and M. Avron: **Photosynthesis 1, Electron Transport and Photophosphorylation.** Encyclopaedia of Plant Physiology, New Series Vol. 5, Springer Verlag 1977.

University of Auckland (1979): **The Potential of Energy Farming for Transport Fuels in New Zealand.** New Zealand Energy Research and Development Committee, Univ. Auckland, August 1979.

University of Sydney (1978): **Alcohol Fuels.** Institution of Chemical Engineers, New South Wales Group, Dept. of Chemical Engineering, Univ. Sydney, Australia, 1978.

UK-ISES (1979): **Biomass for Energy.** Conference (C20) London, UK-ISES, 1979.

Wagner, T. C. (1979), D. S. Gray, B. Y. Zarah and A. A. Kozinski: **Practicality of Alcohols as Motor Fuels.** Society of Automotive Engineers Inc. Warren Dale Pennsylvania, USA, 1979.

Ward, R. L. (1980): "Biological energy conversion: the photosynthetic process", in **Solar Energy Handbook Part A, Engineering Fundamentals.** ed. W. C. Dickinson and P. N. Cheremisinoff, Marcel Dekker, 1980, 563-588.

Watt Committee (1979): **Energy from the Biomass.** Report No. 5, The Watt Committee on Energy Ltd. Birdcage Walk, London, UK, 1979.

White, L. P. (1980), L. T. Plaskett and J. B. Low: "Overview of opportunities for energy from biomass in the European Communities". **Energy from Biomass in Europe.** ed. W. Palz and P. Chartier, Applied Science Publishers, 1980.

World Bank (1980): **Alcohol Production from Biomass in the Developing Countries.** World Bank 1818 H Street, Washington D.C., September 1980.

Chapter 7

WIND

The use of wind energy for powering boats and agricultural machinery has a very long history. There is evidence for wind driven water pumps used by the Persians for irrigation in the fifth century. At the beginning of the fourteenth century the horizontal axis wind turbines mounted on a massive tower which housed a corn mill (whence the name wind mill) became popular in several European countries. They were the main source of mechanical power other than animal or human power for about five hundred years until the invention of the steam engine which was independent of the weather and used energy stored in the fuel itself. The problem of energy storage for wind energy conversion systems has still not been solved although a number of new schemes have been developed in the last few decades.

It seems though that the use of wind energy reappears at the beginning of periods where an increase in energy consumption is envisaged. The need for irrigation in modern, intensive farming made American farmers turn to wind energy and the opportunity to use the same turbines for electricity generation gave extra popularity to these devices. In the USA alone during the first half of the 20th century there were 50,000 units in use until the introduction of electrification into rural communities presented an unbeatable competition and the familiar fan type wind turbines were phased out.

The recent realisation of increased energy consumption caused a revival of interest in wind energy conversion and several countries introduced major national programmes for the development of power plants based on the use of new engineering materials and technological innovations in manufacture, control, aerodynamics, etc. In many industrialised countries wind power will be used to generate electricity and by the year 2000 could make up to a 10% contribution to electricity demand, this figure being limited by the time it necessarily takes to introduce a new technology on a large scale.

Developing countries are also interested, although in their case the use of wind will not be restricted to the generation of electricity but could also make a significant contribution to energy requirements for irrigation and rural development.

7.1 The wind energy resource

The atmosphere of the earth is a large energy converter where amongst other processes solar energy is converted into kinetic energy of the air. Kreith (1978) and Kreider estimate that the earth's atmosphere absorbs approximately 20% of the total incident solar flux. Only a small portion of the energy absorbed by the atmosphere is converted into wind energy and of this only a small fraction can be tapped. It is the intermittent nature of availability and the limitations on the conversion of this energy into power which create all the difficulties and prevent the much wider use of this energy source.

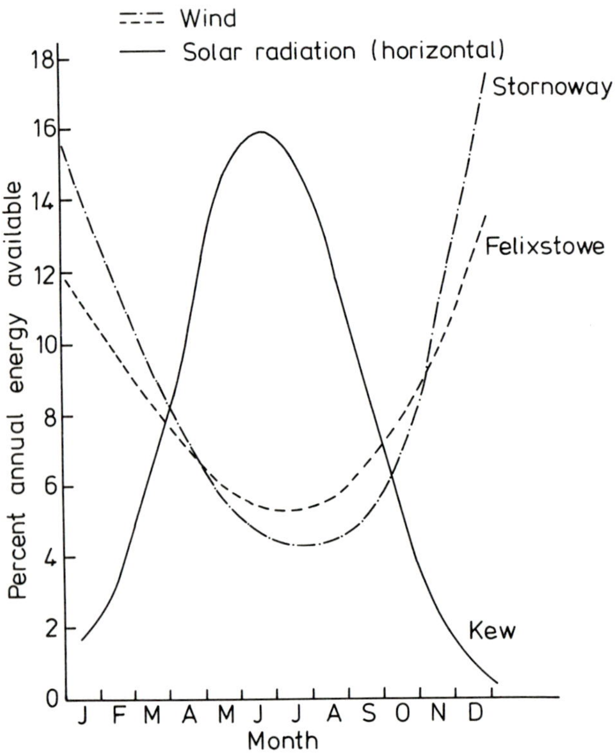

Figure 7.1.1: Wind and solar energy availability distribution.

The energy available from the wind varies with the seasons. In Northern latitudes the peak availability occurs in winter as it is illustrated for two UK stations in Figure 7.1.1. It can be seen that the availability of wind is complementary to that of solar radiation. In a way this opens up some opportunities for combined solar-wind energy utilisation schemes because one can make more use of wind energy in the winter when there is less solar energy available and vice versa. It is not guaranteed however that the two will complement each other well enough to meet demands. This was the case in the project reported by Machens (1980) where the comparison of solar, wind and demand energy distributions made it obvious that the energy storage has to be designed very carefully.

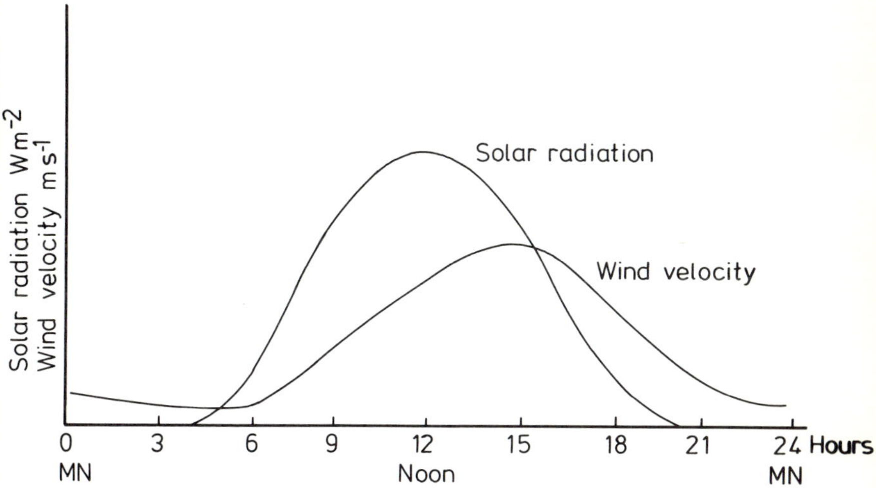

Figure 7.1.2: Diurnal distribution of solar radiation and wind velocity.

The diurnal variations in wind velocity show a different pattern. Here the highest wind velocities occur slightly later than the maximum solar radiation but follow it more or less in the same manner of distribution as shown in Figure 7.1.2. Although this is generally true for the microclimate it is subject to many other conditions of the macroclimate such as pressure distribution around the region in question. It is interesting to compare the diurnal variations of wind power at different sites in the highest and lowest months of availability during the year. Two sites have been selected for this comparison by Allen (1977) and Bird in their report for ETSU: one is Elmdon near Birmingham and the other is Stornoway on the Isle of Lewis. In both cases the winter wind power averaged hourly is higher than the summer wind power for the same hour of the day as shown in Figure 7.1.3. The summer distribution at both stations and the winter distribution at

Elmdon show the usual diurnal variation peaking about two hours after solar noon. The Stornoway winter pattern, however, shows practically no diurnal variation. This is because of the very small solar flux available in the winter at such a high latitude (58°12) and also because the Gulf Stream dominantly determines the wind conditions on this coastal site.

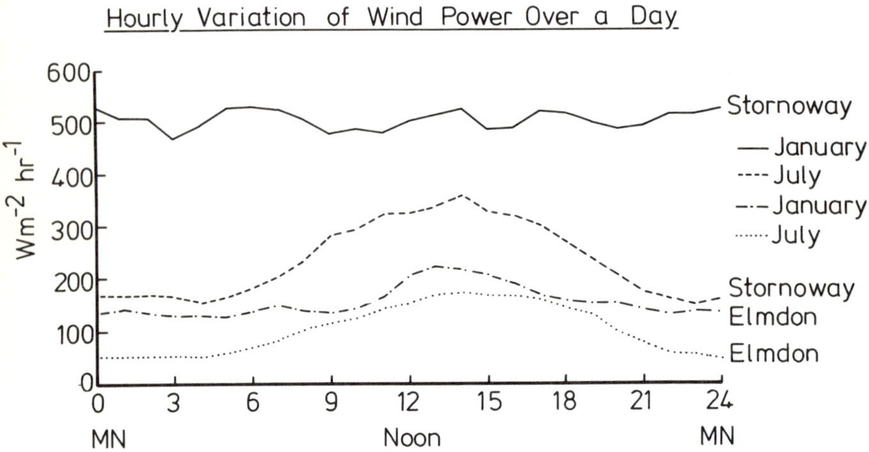

Figure 7.1.3: Diurnal distribution of wind power at Stornoway and Elmdon in winter and summer.

The comparison of available energy densities of solar radiation and wind power indicate that the Birmingham area has similar W/m^2 solar and wind power while the Isle of Lewis has almost four times as much wind power as it has solar power. The Outer Hebrides, the Orkneys, the Shetlands and the southwest shores of the Irish Republic have probably the best wind sites in the world, not so much because of the high velocities experienced there but because the high wind velocities are coupled with long periods of steady high winds. This is desirable because the aerogenerators can be designed for a narrower range of wind speed. Other European wind sites are not as favourable as those on the Scottish Islands. Nevertheless, Denmark, Sweden and Germany have workable areas around the North Sea.

Further inland the yearly average wind speed and the power density drops sharply. Some selected typical values of wind and solar power densities together with wind velocities are given here in Table 7.1.1. The values in this table illustrate the point about the relationship between solar and wind power densities and how much they depend on geographical conditions. It can be seen from this table that most coastal regions have higher wind power densities than inland locations, but there are exceptions like Dakar (coastal) and Amarillo (1000 km from the sea). The highest wind velocities are also found on the sea coast even on off-shore islands like Stornoway (Isle of Lewis) and List (Sylt Island), but there is an interesting exception in the Palm Springs region (200 km from the sea) where a valley connecting the desert with the Los Angeles area acts as a nozzle and accelerates the wind to high velocities and creates a high wind power density. Predictably there are plans to build wind energy converters in the valley.

Table 7.1.1 Comparison of wind and solar energy

Place		Wind		Solar	Wind/Solar
		m/s	kWh/m²	kWh/m²	ratio
UK	Birmingham	3.6	710	950	0.75
UK	Felixstowe	5.1	1600	980	1.63
UK	Plymouth	5.9	3000	1150	2.61
UK	Aberdeen	6.2	3500	830	4.22
UK	Stornoway	7.7	5300	810	6.54
FRG	Karlsruhe	2.4	620	1150	0.54
FRG	Aachen	3.0	890	1050	0.85
FRG	Hamburg	4.3	1500	980	1.53
FRG	List/Sylt	7.2	4620	1000	4.62
USA	Boston area	5.1	1230	1320	0.93
USA	Honolulu/Hawaii	5.2	2100	2190	0.96
USA	Amarillo/Texas	6.1	1930	2000	0.97
USA	Palm Springs/Cal.	9.9	5260	2048	2.57
Senegal	Dakar	4.9	487	2190	0.22
Singapore	Singapore	1.9	75	1533	0.05

As the velocity of the wind changes frequently it is necessary for the designer to know the annual distribution of the various velocities. The velocity duration for Felixstowe and Stornoway are shown in Figure 7.1.4 and it is evident that the wind speed is higher all year round in Stornoway. There are only a few hundred hours during which there is calm for both sites and the yearly mean wind speed for Stornoway is 50% higher than that of Felixstowe. Since wind turbines have to be feathered when wind speeds exceed the design maximum the number of hours of high wind velocity is also a significant parameter.

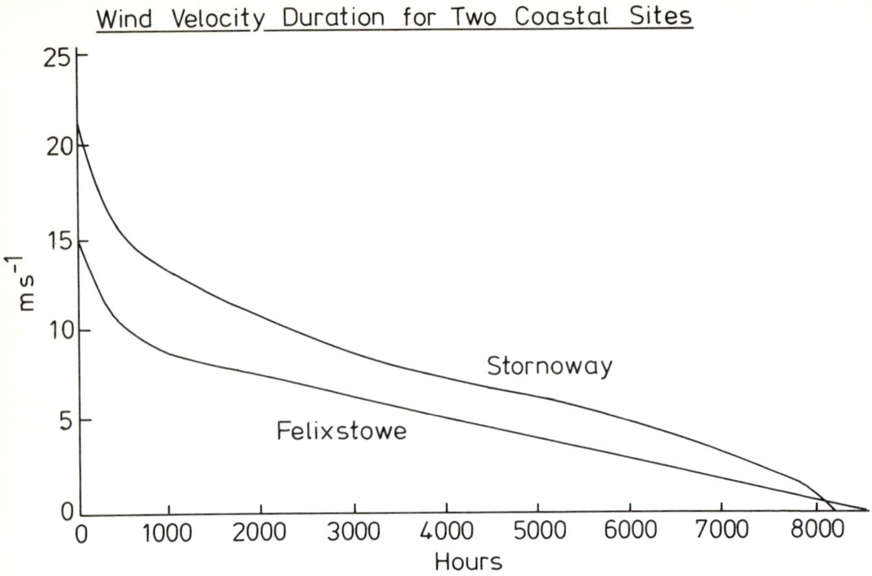

Figure 7.1.4: Annual wind velocity duration curves for Stornoway and Felixstowe.

Another indicator which characterises the site is the power density duration shown in Figure 7.1.5 for the same two locations. Here we can see that the Stornoway site has more than twice the power density when compared to Felixstowe. When the power densities are multiplied with their respective durations the total energy density is 5300 for Stornoway and 1600 for Felixstowe. However, the difference between the two sites in terms of the available useful power is limited by the cut-in and cut-out wind speeds of the turbine and may not be as large as these figures would imply because a large part of the total energy density comes mainly from the high velocity – high power density components.

All this can give an impression of what the designer of a wind turbine installation has to keep in mind, regarding wind velocities and in general the distribution of the available wind resource. A number of studies were made in preparation of various large wind energy projects. In the UK the Central Electricity Generating Board is the major potential customer for wind energy, they are active in a number of wind energy utilisation projects, Moore (1979) analysed wind prediction methods and provided mathematical models for extrapolating on-shore wind data to off-shore sites. Lindley (1980 A) and co-workers have considered the effects of waves and wind on the wind turbine tower structure.

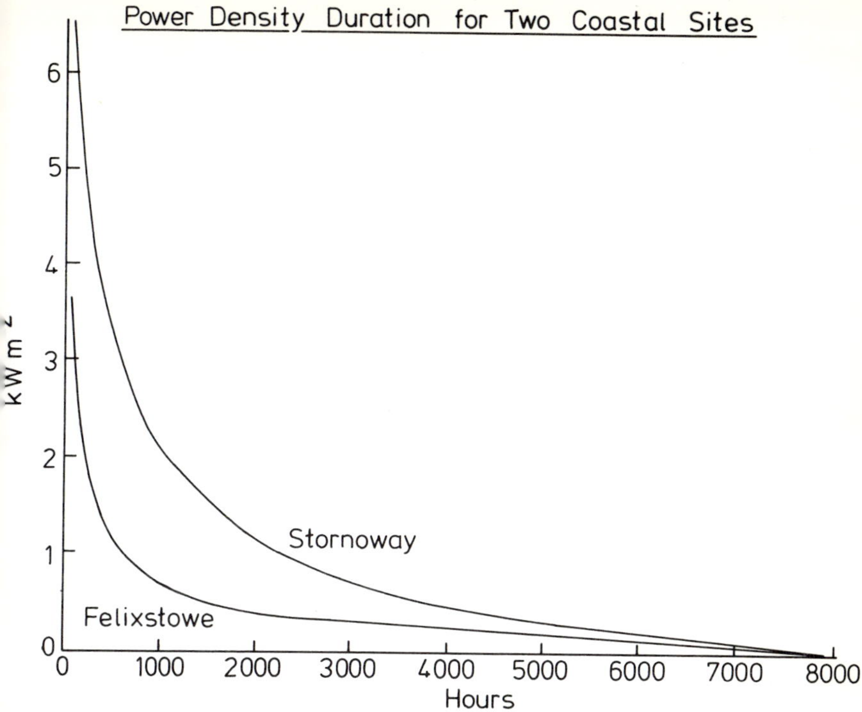

Figure 7.1.5: Annual power density duration curves for Stornoway and Felixstowe.

Lissaman (1980) and co-workers reported on the Palm Springs, California site mentioned earlier. Winds prevailing on the German sea coast are the subject of Duensing's (1980) report, which compared wind velocities both in the horizontal distribution towards the inland and in the vertical distribution normal to the surface.

A new approach to wind speed characterisation by the use of time series analysis and modelling was reported by Goh (1979) and Nathan. They compared two inland, two coastal and two off-shore wind station data in Singapore, spread over an area of 700 m^2. Further results of their studies were described in their 1981 paper at the British Wind Energy Association's International Colloquium on Wind Energy which was held in Brighton, UK, where they suggested methods for evaluation of stability of wind characteristics from year to year.

An interesting solution to the problem of obtaining adequate site-specific wind data was found by the Oregon Department of Energy, Philbrick

(1979) and Morgan described a wind anemometer loan programme which has motivated local initiatives in various communities for the assessment and utilisation of wind power.

When the results of a survey are obtained the data have to be analysed. There are many approaches and one that is frequently used is modelling through probability distributions. Examples of this can be found in the literature and we refer to Justus (1978) and co-workers' method for estimating wind speed frequency distributions and to Swift-Hook (1979 A) who reviewed methods of presenting wind energy data. An international collaborative programme of field measurement on wind turbines involving a wind survey and on-site analysis of wind turbine performance was reported by Swift-Hook (1979 B) during the first BWEA Workshop.

Finally it is worth noting that the information on the wind energy resource is seldom available in one publication. Some references are given in Chapter 1. Recently as a result of an International Energy Agency project detailed wind data for West Germany has been published by Jarass (1981) and co-workers. An interesting example of a compilation of solar and wind energy data for one region is that of Falicoff (1979) and co-workers for Hawaii.

7.2 Wind energy conversion concepts

The wind blows horizontally most of the time and to make use of the kinetic energy in the moving air it is natural to think of some device which will bring the air to rest so that it gives its kinetic energy to the device. Sails on boats which move with the wind extract part of the kinetic energy but even with a stationary wind machine it is not possible to convert all the kinetic energy in the wind into mechanical power.

If, however, a wind machine is designed in such a way that it allows some of the air mass to move on while some is slowed down, the pressure created in front of the machine can be released in a continuous fashion and a sizeable part of the kinetic energy can be converted by the machine. The sails of the traditional Dutch windmill are set at an angle which allows the air to pass through the rotor between the sails while the sails are pushed sideways in the circumferential direction, thus creating a rotary motion. For a particular range of wind speeds it is possible to select the right ratio between the area of sails and the total area of the rotor and by selecting the correct angle between the sails and the oncoming wind the windmill will keep turning as long as the wind blows. It is also necessary to ensure that the rotor always faces the wind.

The amount of power that such a windmill can develop will depend on a number of characteristics, especially the diameter of the rotor. The speed with which the rotor turns will depend on such factors as the wind velocity on the angle to which the sails are set and the load applied to the driven shaft.

The traditional Dutch windmills and the American farm windmills are mainly used for agricultural purposes: water pumping and corn grinding. As the availability of the wind cannot be guaranteed or controlled it has to be utilised for jobs which are not critically dependent on time. Water pumping is like that and provided the corn is all prepared and waiting to be ground it also can be done at any time. Thus these two types of windmill have developed in perfect harmony with nature and traditional agricultural practices without the need for energy storage.

The power needs of high intensity farming, industry, commerce and modern life in general are such that they could not be supplied by those traditional windmills. Modern wind machines are designed with much greater efficiency and to generate electricity.

Unlike their agricultural predecessors the modern horizontal axis machines use only two or three slender blades attached to the hub which is mounted on a shaft driving the generator through a gearbox in the nacelle on top of a slender tower. The direction of the wind is sensed by instruments and an electric motor drives the nacelle to a position where the rotor is facing the wind. This can also be helped by tilting the rotor blades out of the plane of rotation (coning) so that the wind itself orients the rotor. This method reduces the performance for a given blade lebgth because the rotor diameter is reduced by the coning. The wind speed is also measured and the pitch angle at the root of the blades can be adjusted accordingly to optimise the aerodynamic performance. The best angular setting requires many calculations and the whole wind turbine is controlled by a computer, often placed in the nacelle.

These new developments have changed the visual appearance of wind turbines and created a new application of wind energy conversion: the large scale generation of electricity by wind power. It has also meant that in addition to individual users who had a local need for small amounts of power which could be supplied by the wind the national power generating companies have become interested. They have a ready made market, the distribution network and the necessary capital to make large wind turbine projects possible. To them the wind is just another energy source and the use they make of it will depend on the costs and the ability to match the pattern of electricity use. This matching of the supply and demand can bring in the question of storage and in any case has a great influence on the operating efficiencies of the various components of the power generating system.

Augmentation by specially designed surfaces which lead and accelerate the wind is the subject of research and development in several countries. In Israel a shroud is used by Igra (1980) and Schulgasser to improve the aerodynamic efficiency of the turbine, in much the same way as it is done on aircraft jet engines.During recent tests the unit which is 8m long with a 3m diameter throat and exit diameter of 6 m produced one horsepower from the 3 m diameter axial flow horizontal axis twelve bladed turbine rotor at a free stream velocity of 5 m/s wind. This is more than one could obtain from an open rotor turbine of the same diameter, but without the shroud. The power augmentation factor of two is claimed by the inventor for this turbine shown in Figure 7.2.1.

A much larger power augmentation factor is claimed by Greff (1981) and Holzdeppe on a different design. Working at the Institute for Aerospace Engineering in Aachen, they developed a wind energy concentrator which consists of a cambered delta wing surface mounted on a turntable. The adjustable attitude angle helps to control the generation of vortex by the control of the angle of attack. The delta wing shape lends itself to a symmetric double arrangement where there are two turbines one at either side of the symmetry axis of the vortex generator. The two vortices are intercepted by the turbines near the trailing edge of the delta wing vortex generator. Wind tunnel tests showed a ten fold power augmentation factor, according to the authors, who report that field testing is currently under preparation. Although the initial capital investment costs are high, the authors estimate that the installation is going to be cost-effective. The same basic idea has been reported by Sforza (1977) who worked on vortex augmentation in the Aerodynamics Laboratories of the Polytechnic Institute of New York for several years.

Ducted rotors arranged in a bank to form a "wind wall" have been proposed by the Environmental Design Group for a housing project in

Figure 7.2.1: Shrouded wind turbine at Beer Sheva University.

Sussex. McVeigh (1977) describes this proposal which combines solar collectors underneath the wind wall with augmented wind power. The designers claim a wind energy conversion efficiency of 77% which means perhaps a factor of 2 in power augmentation. The scheme has drawbacks: the ducted turbines are fixed and unable to turn into the wind; also: the flat plate solar collectors will have higher heat losses if they are used for channelling the accelerating wind into the ducted turbines.

The combination of augmented wind and solar power in one structure has been tried before. Architect Maloolm B. Wells designed a house which was built in Raven Rocks, Ohio, USA. It features a funnelled wind energy collector on the roof of the building. Anderson (1977) describes how the sloping walls increase the wind speed, but it appears that the major features of the house design are centered around passive solar energy use and the wind power augmentation is of less importance.

7.3 Horizontal axis wind turbines

The traditional windmills had four sails in their rotor and the rotor was mounted on a hub which was attached to the roof of a massive building which acted as a tower. The roof structure was a turntable and most of the moving parts were made of wood. The sail area could be made smaller by furling the sail cloth or as in some designs by adjusting the louvre-like wooden slats. This was necessary when the wind speed was too high for the best operation of the mill. The rotor could be turned into the wind by a tail fan which in itself was a small wind turbine.

Although the romantic public image of these windmills links them to Holland, they have been used in many countries for hundreds of years. From early mediaeval times the landscape of Germany was dotted with windmills, and we know of windmills in Spain where the valiant Don Quijote was reported to have attacked at least one. In France the writer Alphonse Daudet lived and wrote his 'Lettres de mon moulin' in a windmill, a fashionable lifestyle which found many followers later in England. In East Anglia, which has a geography not unlike that of Holland, there are still hundreds of old windmills although many of them are now in ruin.

Figure 7.3.1: Single bladed wind turbine with counterweight.

The rotor of horizontal axis wind turbines can have any number of blades, the current thinking is that two bladed rotors are the most practical solution as they are well balanced and represent a small solidity ratio which is desirable for large installations. Medium size machines often have 3 bladed rotors and only small turbines have rotors with more than three blades. Some people believe in using a single blade with a counterweight for balance, as shown in Figure 7.3.1 but this design has not become as popular as the two bladed rotors Figure 7.3.2 or the three bladed rotors Figure 7.3.3. The multibladed rotor was once very popular but is now quite rare.

Figure 7.3.2: Two bladed wind turbine on the island of Pellworm, Germany. Power rating is 10kW in 8 m/s wind.

Figure 7.3.3: Three bladed rotor of a 10kW wind turbine.

An historic step in the development of the modern wind turbine, was the construction of a wind turbine on Grandpa's Knob near Rutland in Vermont, USA, in 1941. It had a power capacity of 1250 kilowatt developed by a two bladed rotor of 53 m diameter mounted on top of a 33 m steel tower. After years of careful preparation and analysis the site was selected because its long term wind characteristics were favourable and the designers wanted to make use of the natural augmentation which results from the wind accelerating up the slope. The sheer size of the turbine and boldness of the design captivated the imagination of wind energy enthusiasts all over the world for decades to come. In March 1945 one of the blades, weighing 8 tons, broke off and flew 230 meters before it landed. At the time it was not appreciated that the damage had been caused by dynamic loading as opposed to static loading. Although there were other large machines built before this American one, such as the German Kumme design in 1920, the French Darrieus machine with its two bladed 20 m rotor built in 1929 and the Russian 30 m rotor, 100 kilowatt machine erected near Yalta in 1931, the so-called Putnam turbine on Grandpa's Knob is the most famous of them all. An interesting account of its history was given by Putnam (1948) the project manager. He received his inspiration in 1934 when he built a summer house on Cape Cod in New England where he "found both the winds and the electric rates surprisingly high". His book remains a popular favourite on the subject and in 1981 he received honours from the wind energy community for his work.

Horizontal axis machines can have their rotor upwind or downwind, relative to the tower. The turbulent wake created by the tower causes cyclic pressure variations which increase fatigue problems in the blades. Jesch (1981) and co-workers at Birmingham University measured these effects in wind tunnel tests and found that the pressure affects the blade even in the upwind rotor position. Unsteady aerodynamic forces acting on the blades were studied by Duremberg (1979) at Exeter University. The solution to these problems lies in good design and good manufacturing techniques. Herepath (1979) and co-workers at University College, Swansea, described the manufacturing method of a 5 m diameter rotor with blades having variable setting angle. Danish engineers use fibreglass and carbon fibre technology, based on boat building technology whilst German workers use aircraft construction methods to build rotor blades. The giant Growian I, the largest machine in Europe has a 100.4 m diameter rotor. The steel blades are 4.9 m wide and 2.4 m thick at the root and they weigh 23 tons each. The project was described by Huss (1980) with special attention to the aerodynamic design of the rotor. Although the size of the rotor presents special problems to the blade design it seems that there is no unique material or construction method.

As a rough approximation one can classify the horizontal axis wind turbines into three categories: 10 kW, 100 kW and 1000 kW units. The smallest of these is usually a machine with about 8-10 m rotor diameter and costing a little more than an automobile. Almost all the available units are developed by private manufacturers and a great variety exists. The medium category machines may have rotors 15-30 m in diameter and only large industrial companies are involved in their development. They usually receive support from national governments or are developed through cooperative company projects, sometimes involving companies from several countries. The largest units are in the megawatt range and typically have 60-100 m diameter rotors. They are funded at the national and international level. The projects which produce such huge machines are often beset with problems but the economy of scale is in their favour. In a recent paper Lindley (1980 B) gives a survey of current projects in Europe and in the USA.

The examples of the small machine category are numerous. They are manufactured in the United States, Germany, Denmark, Sweden, United Kingdom, France, Italy, Switzerland, Holland, Hungary and the Soviet Union. The large variety of designs, construction and performance makes the choice for the user a difficult task. In an attempt to give a fair evaluation under identical wind conditions the leading nations have established proving grounds for the comparison of different designs. The US Department of Energy's Rocky Flats test center in Colorado can evaluate up to 30 turbines simultaneously. Divone (1981) reported on the organisational flexibility of the center and its role in providing independent assessment of the different commercially available units. To help with the development of small systems the Federal Government of Germany established a test site on Pellworm Island off the Frisian coast. Fries (1980) and co-workers reported on the wind survey aspects of the test site and Fritzsche (1981) analysed the importance of the test programme carried out at the site and the role it plays in the national energy strategies of developing countries who come there to select suitable equipment. On this site machines of different rotor configurations, all having a nominal power rating of 10 kW in 8 m/s wind are being evaluated under identical conditions. The Danish test site at Riso has a similar function and although its wind characteristics are less favourable, it benefits from the excellent supervision it receives from the technical and scientific staff of the nearby University and Government Research Centre.

Small machines are popular in remote areas where electricity is not available and the cost of oil for power generation is prohibitive. As they require energy storage the load matching with power availability can be a problem. An excellent application is water pumping for irrigation of farm lands, because the storage is represented by the height of the water level in the distributing reservoirs and it can be independent on the discharge rate into the irrigation canals. Sometimes the energy is stored in thermal loads of various kinds but small machines are not suited for high efficiency solutions of thisproblem. A survey by Kharitonov (1981) describes the use of small wind turbines in the Soviet Union.

The aerodynamics of small machines have received much attention in recent years. The choices available for the designer of high performance blades were investigated by Jesch (1979) and Yeo. Aerodynamic considerations, stress analysis and constructional techniques used in the design and manufacture of a 5 m diameter rotor with variable setting angle was reported by Herepath (1979) and co-workers. A reduction in cost by using a simplified linearised blade design was reported by Jesch (1980) who used this method in the making of GRP blades for a 5 m diameter rotor in Birmingham.

The working examples of medium size horizontal axis wind turbines are less numerous but more interesting because they operate in a category where direct coupling of the output with the local or national grid is possible and the fuel savings are potentially greater. One of the early British projects was concerned with the design, construction and installation of a three bladed rotor capable of developing 100 kW. The John Brown

Figure 7.3.4: British wind turbine near Boroughbridge, North Yorkshire.

company produced the prototype and it went into operation in 1950 on Costa Hill in the Orkneys. The wind survey of the site was very thorough and it is described by Golding (1976) in the updated version of his 1955 classical book on the generation of electricity by wind power. It is interesting to note that 31 years later a new effort is being made not only to build prototype machines in the Orkneys but also to provide electrical power to the local community. On a site about ten miles away from Costa Hill there are two units being designed, built and tested by the newly formed Wind Energy Group, comprising Taylor Woodrow Construction Ltd., British Aerospace Public Ltd. Company and G.E.C. Power Engineering Ltd. The first machine is 20 m in diameter and it is funded by the North Scotland Hydro-Electric Board together with the UK Department of Industry and the Wind Energy Group. Lindley (1981) and Stevenson reported that lessons learned from the design, construction and operation of this machine will be utilised in the second project which is to be funded by the Utility Board and the UK Department of Energy. This project will produce a megawatt class machine.

Another UK achievement reported by Nickols (1979) was the medium size wind turbine near Boroughbridge, shown in Figure 7.3.4, it was the result of a private initiative by Sir Henry Lawson-Tancred. It has produced electricity since 1976 and has used a direct hook-up to the national grid. The problems of integrating wind turbines with electricity supply grids were investigated by Lipman (1980) and co-workers at Reading University while Diesendorf (1981) and co-workers working in Australia have analysed the economic value of wind power to the electricity grids.

The top end of the medium size machines is well represented by the example of the two Danish machines installed and operated at Nibe and shown in Figure 7.5.5A. The 40 m diameter rotor develops 700 kW power for the national grid. Pedersen (1980) and Nielsen described the project recently and the interesting solution of the blade construction is shown in Figure 7.3.5.B where the D shaped spar can be seen. The blade is made of fibre glass, carbon fibre and steel. It transmits the torque to the shaft through a steel plug which fits into the D shaped spar, which itself is wound from continuous carbon fibre filament and reinforced resins. One of the most outstanding characteristics of the Nibe machines, as well as the whole of the Danish programme is the low cost of the installation. This is partly due to the short project time and partly to the judicious choice of size. Building two 700 kW machines quickly and at low cost not only produces the desired power generating capacity but it also provides operating experience and rapid feed-back of information for the next design.

Figure 7.3.5.A: Nibe B abd Nibe A, the two most recently installed wind turbines in Denmark. *(Courtesy Ministry of Energy and Electric Utilities in Denmark).*

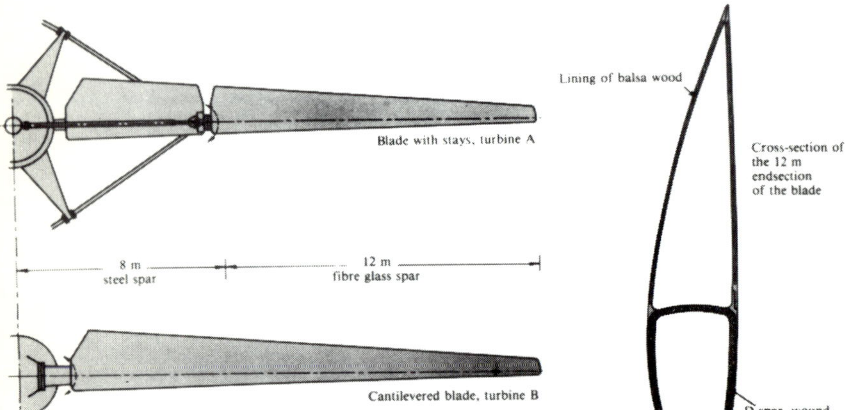

Figure 7.3.5.B: Rotor blade details of Nibe A and Nibe B machines. *(After the official publication).*

Other medium scale projects have been reported by Spittler (1980) who described a soft and slender blade design under construction in Germany. Conceived by Hutter it develops 265 kW with its slowly rotating flexible blades. Under wind pressure the blades bend backward so much that a special tower design was needed. In the United States private industry is becoming interested in producing medium size machines. A prototype unit installed on the offshore Cuttyhunk Island has a 26 m diameter rotor and it develops 200 kW in 13 m/s wind. The operating experience was good enough to convince the utility companies of neighbouring Southeast Canada and Northeast USA to purchase two production units. Recent trends however point in the direction of the viability of megawatt size machines.

Examples of large aerogenerators were few in the past fifty years but in the last five years they have started to operate in many places. The Soviet Union is reported to be working on large machines and Hugosson (1981) believes that they may be tested in the Ural mountains. Specific information on the Soviet programme is hard to come by. The United States built the first megawatt unit in 1941 but three decades passed before the new generation of machines appeared both in the coastal regions and inland. West Germany is involved in the Growian projects and also in cooperative enterprises with Sweden. American and German industry are working with Swedish industry and Government on two large aerogenerator projects. Britain is set on course to build the 60 m diameter wind turbine on Costa Hill and make use of the experience gained from the 20 meter machine. So the scene in Europe and on the American continent is busy with activity. With Government support the engineering development done by private industry on big machines is coming to fruition and prototypes are giving way to working units.

Having learned the lessons from the so-called Mod-O wind turbine, first installed in 1975 in Sandusky, Ohio, four modified versions with twice the output were installed in Clayton, New Mexico, Culebra, Puerto Rico, Block Island, Rhode Island and Kuhuku Point, Oahu Island in Hawaii. The selection of these sites were motivated by earlier indications of suitable wind conditions and were based on extensive site surveys.

Soon after the first generation of large wind turbines, the Mod-1 series followed. The first one was built in Boone, North Carolina, and it was the first megawatt size machine in 35 years. Also for the first time a long run of successful operations followed at full rated power with no difficulties in frequency control, voltage control and system reliability. New problems were encountered however: TV signal interference and low frequency, low level noise caused by the blades passing through the wind shadow of the tower. The introduction of cable TV solved the first one and a change in the operating rotational speed ameliorated the second.

The scene was set to the current phase in the American programme: the Mod-2 aerogenerators with their 91.5 m diameter 2 bladed rotor producing 2.5 MW in 9 m/s wind were installed. As can be seen in Figure 7.3.6 a cluster of three big machines were installed near Goldendale in the state of Washington. It is a significant new development as it will enable the operators to evaluate the behaviour of clusters. It seems inevitable that wind turbines have to be sited in groups concentrated on relatively small land if they are to become significant energy producers for the electrical network.

Figure 7.3.6: The first ever megawatt cluster operating in the US since 1981.

The problems of wind turbine clusters were the subject of investigations by several people in the UK. Lipman and his co-workers at Reading University investigated cluster efficiencies. Bossanyi (1980) and co-workers considered interactions between units in a cluster. Swift-Hook (1979) and Taylor have analysed directional arrays of windmills from the standpoint of the Central Electricity Generating Board. Railly (1981) at Birmingham University calculated the limitations of available power in wind turbine clusters. These analyses will be helpful when the UK programme reaches the stage of building clusters either on land or off shore in the near future.

The growth in size has not stopped with the American Mod-2 machines. In West Germany the Growian I is being built which is not only the largest machine in Europe but it is also the largest in the world. The construction of the 100.4 m diameter rotor was described by Thiele (1980) and the report of Huss (1980) gave an account of the project.

Occasionally it is a grass roots movement which produces interesting results. Two high schools and a teachers' training college initiated a wind energy project in 1974 in Denmark. In three years they succeeded in designing and building a horizontal axis turbine with a 54 m diameter rotor which became known as Twindmill. The unit produces 2 MW power in 15 m/s wind which is used mostly to preheat a well insulated thermal store of 3000 m^3 water tank which serves the community. The surplus output is fed into the national grid. Kalke (1977) gave an early account of the project which was sustained by largely voluntary contributions from amateurs and professionals.

Finally an unusual design should be mentioned: the single bladed rotor. Here the blade is counter-balanced by weight on a short arm. The aerodynamic efficiency increases and the proponents of the idea say that this alone is worth all the drawbacks presented by the design. The design concept is illustrated on a smaller experimental model shown in Figure 7.3.1. The first rotation of a 48 m diameter prototype was in 1981 near Bremerhaven in Germany. The development of the full size unit with its 145 m diameter rotor is less advanced, but according to Meggle (1980) and Schobe it is expected to have a power output of 5 MW at 11.4 m/s wind velocity. This is about the size of a small thermal power station and the success eventually might lead to clusters which can supply the electricity needs of the nearby community without using imported oil or domestic fossil fuels.

7.4 **Vertical axis wind turbines**

By making the axis vertical the rotor will always be "facing" the wind removing the necessity to turn the rotor into the wind. A further advantage of these machines is that the drive mechanism can be placed at the bottom of the axis, thereby lowering the centre of gravity for the whole system and making it more stable. The disadvantages are in the lower efficiency of the overall system and the difficulty of anchoring the top of the vertical axis to the ground. Additionally the centre height of the rotor is also lower than that of similar size horizontal axis wind turbines which means that the average wind velocity reaching the rotor is also less due to boundary layer effects. The graph in Figure 7.4.1 compares the efficiency of various types of rotors with the ideal, maximum efficiency possible. The efficiencies are plotted against the tip speed ratio which expresses the relationship between the circumferential velocity of the blade and the oncoming wind speed.

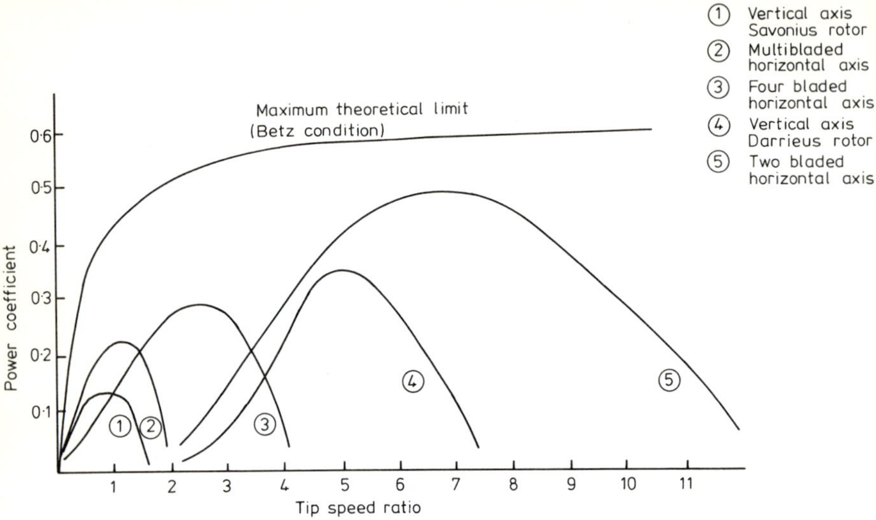

Figure 7.4.1: Comparison of rotor efficiencies as functions of tip speed ratio.

A simple inspection shows that:

a. Each rotor type has a characteristic tip speed ratio at which it reaches maximum power coefficient and that these vary with different designs.

b. there is a general tendency of higher tip speed ratios yielding higher power coefficients.

c. although the two bladed horizontal axis machines being developed in the 1980's give the highest power coefficient, the vertical axis machines are the next highest while the Savonius rotor and the multibladed horizontal axis machines have the lowest power coefficient.

d. the tolerance of the rotor to changes in the dimensionless tip speed ratio is greatest in the newly designed two bladed horizontal axis rotors and it tends to narrow at lower tip speed ratios.

A certain value of tip speed ratio can be achieved by either controlling the rotational speed or the wind speed. Clearly only the first is possible. The rotational speed which gives the circumferential velocity as the product of radius and angular velocity offers two choices. Either small radius and high angular velocity or large radius and low angular velocity will give the same tip speed ratio for a given available wind velocity. The reason why modern machines use the latter is that in addition to the economy of scale they also can catch higher wind velocity with a larger rotor radius because they intercept wind at higher altitudes.

Vertical axis wind turbines of current design seem to operate best at around a tip speed ratio of 5. This classifies them as high speed machines as the comparison is always made with reference to earlier machines, which had tip speed ratios between 1 and 3.

Figure 7.4.2: The 17 m diameter Darriens rotor at Albequerque, New Mexico, in 1979.

The most significant modern invention in the application of vertical axis wind conversion principles was made by the French inventor Darrieus in the 1920's. Ignored largely for some fifty years his idea was rediscovered and championed by Templin (1974) in Canada. The Darrieus rotor has a unique shape, often called the eggbeater. As shown in Figure 7.4.2 the blades are catenary which is the line of a rope which is spun around an imaginary vertical axis. Thus the blades of the Darrieus rotor are preformed to the shape which has the minimum stresses in operation. The aerodynamics of the Darrieus rotor were investigated in the last five years in Britain by Sharpe (1976 and 1978), Jesch (1978, 1979, 1980) and Walton, Read (1980) and Sharpe as well as by Jesch (1980) and Crossland.

Although much has been learned from these studies there is yet no industry in Britain to produce these machines on a large scale. Watson (1979) and Bainbridge reported the design and construction of a 4.5 m diameter test unit which was first operated in Newcastle in 1977. Watson (1979) also looked at the self starting capabilities of low solidity fixed pitch Darrieus rotors and recently Jesch (1981) and Walton have analysed the cyclic torque variations in Darrieus rotors and gave an explanation for spontaneous self start, which they verified in wind tunnel tests.

In Europe the Dornier company is leading in the production of Darrieus type turbines. Fritzsche (1980) and Wirths reported on a cooperative venture with Argentina. The first unit has Savonius type S shaped rotors mounted on the vertical axis to help start the turbine in Patagonia since 1981. In Sweden a new conceptual version of the Darrieus principle was proposed by Ljungstrom (1980) which uses an essentially single bladed design. A small scale model of the multimegawatt machine has been built and is being tested.

Figure 7.4.3: Savonius type rotors help to start this Darrieus rotor made by the Dornier company in 1980.

In Canada medium size, high reliability machines have been mass produced recently. In spite of earlier teething problems and the much publicised spectacular failure of the Magdalen Island installation caused by the maintenance crew not applying the brakes on the rotor at the end of the day resulting in selfstart which was previously thought impossible the Government and the Hydro-Quebec utility company are building a 3 MW system. When completed in 1984 it will be the world's first megawatt size Darrieus machine.

Three machines were built and operated as test prototypes in Albuquerque, New Mexico, by the Sandia Laboratories. The unit shown in Figure 7.4.2 was completed in 1977. Experience with helicopter blade technology helped in the development of this 17 m diameter rotor. A third blade lying on the ground has 0.6 m chord length and weighs two tons. It could be installed to form a three bladed rotor for the purpose of studying the effect of the number of blades on maximum power coefficient. The largest American Darrieus rotor, shown in Figure 7.4.4, is operated since 1980 in a three bladed version and it develops 500 kW power. In recent years there has been a new interest in building larger Darrieus rotors and the UK Department of Energy has studied the case according to Allen (1977) and Bird who reported on the potential contribution wind can make to the economy. No large unit has been built yet in the UK to date.

Figure 7.4.4: The largest Darrieus rotor in 1980.

7.5 **Vertical axis straight bladed wind turbines**

The analysis of the Darrieus rotor's performance shows that the equatorial part of the blade contributes much more useful torque than the parts which are closer to the poles. This is due partly to the larger radius and partly to the angle of the blade portion. This led to the suggestion that straight bladed vertical axis wind turbines would perform better because they benefit from the maximal equatorial radius and the angular position as well.

Figure 7.5.1: Experimental unit at the Reading test site.

Musgrove (1976) working at Reading University built such a machine. As shown in Figure 7.5.1 the rotor was an H shape and the central horizontal bar supports the two vertical straight blades. Other rotors were also built and Musgrove (1978) and Mays reported new design features of variable geometry. By tilting the blades from an H shape to a V and later to an A shape the response of the rotor to variable wind velocities could be better controlled.

The aerodynamics of straight bladed vertical axis turbines have been studied by Bannister (1979), and Willmer (1979) reported on wind tunnel tests on the blades. The first application came when the Rutherford Laboratory built the turbine for their Autarkic House. Littler (1979) and Thomas reported on the project which was later abandoned apart from the work on the straight bladed vertical axis wind turbine.

There is interest in exporting such machines to developing countries. An industrial consortium was set up from Sir Robert McAlpine & Sons Ltd., Aircraft Designs (Bembridge) Ltd., N.E.I. Clark Chapman Cranes Ltd. and Engineering and Power Development Consultants Ltd. Their first model is shown in Application 8.36 with its new control mechanism which gives variable geometry. Clare (1981) and Allan reported on the current study which is part of a five year plan aimed at developing larger offshore machines of this type.

A somewhat different design was adopted by the German firm ERNO which has long established experience in aircraft building. Zelck (1980) reported on a 28 m diameter rotor with 14 m long blades made of mahogany plywood which has passed fatigue testing already. The computer simulation of the scale model, according to Scholz (1980) and Simhan, it will be used on the large machine. It will be the first straight bladed vertical axis machine in the 100 kW class.

7.6 Concluding remarks

The conversion of wind energy into electric power serves an immediate purpose and provides savings in fossil fuels. The difficulties caused by TV interference, noise and visual impact of wind turbine clusters have been recognised and efforts are being made to overcome them. Wind technology is developing very rapidly and with the advent of megawatt size machines it is not too far off before reaching a mass production capability. It is clear that whatever contribution wind power can make it could be fully developed before the forecasted decline in the oil supply. The natural development of the new technology is such that a variety of sizes will be available simultaneously by the end of the 1980's thus providing a flexible response to the differing needs of society.

References

Allen, J. (1977) and R. A. Bird: **The Prospects for the Generation of Electricity from Wind Energy in the United Kingdom.** Dept. of Energy, Energy Paper No. 21, HMSO 1977, 67 p.

Anderson, B. (1977): **Solar Energy: Fundamentals in Building Design.** McGraw-Hill 1977, 374 p.

Bannister, W. S. (1979): "Aerodynamic studies of a straight bladed vertical axis wind turbine". **Proc. of the 1st BWEA Wind Energy Workshop,** Apr. 1979. Multi-Science Publ. Co. 88-101.

Bossanyi, E. A. (1980), C. Maclean, G. E. Whittle, P. D. Dunn, N. H. Lipman and P. J. Musgrove: "The efficiency of wind turbine clusters". Papers presented at the **3rd Int. Symp. on Wind Energy Systems** held in Copenhagen, Aug. 1980. BHRA Fluid Engineering, 401-416.

Clare, R. (1981) and J. Allen: "Progress with the Design of a 25 metre diameter vertical axis wind turbine". **Proceedings of the third BWEA Wind Energy Conference.** Cranfield, April 1981. BHRA Fluid Engineering, 33-39.

Diesendorf, M. (1981), B. Martin and J. Carlin: "The economic value of wind power in electricity grids". Paper presented at the **International Colloquium on Wind Energy,** Brighton 1981, BWEA.

Divone, L. V. (1981): "Wind energy development in North America". Paper presented at the **International Colloquium on Wind Energy,** Brighton 1981, British Wind Energy Association.

Duensing, G. (1980): "Das regionale Windangebot an den deutschen Küsten", in **Tagungsbericht des 3 Internationalen Sonnenforums – Vorträge,** Hamburg 1980, DGS-Sonnenenergie Verlag, 441-449.

Duremberg, C. J. (1979): "Unsteady aerodynamics of vertical axis wind turbines". **Proc. of the 1st BWEA Wind Energy Workshop,** Apr. 1979, Multi Science Publ. Co., 117-124.

Falicoff, W. (1979), G. Koide and P. Takahashi: Solar/Wind Handbook for Hawaii: **Technical Applications for Hawaii, the Pacific Basin and Sites Worldwide with Similar Climatic Conditions.** US Department of Energy, Contract #2469609, University of Hawaii and State Energy Office, Hawaii, May 1979.

Fries, S. (1980), D. Borchers and G. Petersen: "Vergleichende Untersuchungen des, Betriebverhaltens von Windenergieanlangen der Leistungsklasse 10 kW" in **Statusbericht Windenergie** VDI Verlag, 1980, 341-354.

Fries, S. (1981), G. Petersen and J. Mohn: "Sea water desalination powered by wind energy". Paper presented at the **International Colloquium on Wind Energy,** Brighton 1981, British Wind Energy Association.

Fritzsche, A. (1980) and G. Wirths: "Wind energy converter with vertical rotation axis". in **Solar Energy in the 80s,** Conference (M2) Birmingham, UK-ISES Midlands Branch 1980, 16-21.

Fritzsche, A. (1981): "Chancen der Windenergienutzung." **Naturwissenschaften, 68,** 1981, 157-162.

Goh, T. N. (1979) and G. K. Nathan: "A new approach of wind speed characterization for wind energy studies". Proc. ISES Silver Jubilee Congress in Atlanta, 1979, Vol. 3, 2267-2271.

Goh, T. N. (1981) and G. K. Nathan: "Stability analysis of annual wind speed characteristics". Paper presented at the **International Colloquium on Wind Energy,** Brighton 1981, British Wind Energy Association.

Golding, E. W. (1976): **The Generation of Electricity by Wind Power.** E. & F. Spon, 1955, reprinted 1976.

Greff, E. (1981) and D. Holzdeppe: "Wind energy concentration within vortex flowfields and its utilization for the generation of energy". Paper presented at the **International Colloquium on Wind Energy,** Brighton 1981, British Wind Energy Association.

Herepath, R. G. (1979), M. G. Wollard and R. T. Griffiths: "Blade design and construction for a horizontal axis wind turbine". **Proc. of the 1st BWEA Wind Energy Workshop,** Apr. 1979, Multi Science Publ. Co. 68-77.

Hugosson, S. (1981): "The European windpower scene". **Proc. of the 3rd BWEA Wind Energy Workshop,** April 1981.

Huss, G. (1980): "Aerodynamische Rotorauslegung der grossen Windkraftanlage Growian". in **Tagungsbericht des 3. Internationalen Sonnenforums - Freie Vorträge -** Hamburg 1980, DGS-Sonnenenergie Verlag, 406-415.

Igra, O. (1980) and K. Schulgasser: "Design and construction of a pilot plant for a shrouded wind turbine". Papers presented at the **2nd Int. Symp. on Wind Energy Systems** held in Amsterdam, Oct 1978, BHRA Fluid Engineering, Vol 1, F1-12.

Jarass, L. (1981), L. Hoffman, A. Jarass and G. Obermair: **Wind Energy** (An assessment of the technical and economic potential. A case study for the Federal Republic of Germany, commissioned by the International Energy Agency.) Springer Verlag, 1981.

Jesch, L. F. (1978) and D. Walton: "Design analysis of a vertical axis wind turbine". 2nd International Solar Forum, Hamburg 1978, Vol. 3, 157-168.

Jesch, L. F. (1979) and D. Walton: "Performance modelling of the vertical axis wind turbine". Proc. ISES Silver Jubilee Congress in Atlanta 1979, Vol. 3, 2277-2281.

Jesch, L. F. (1979) and A. T. Yeo: "Design and testing of quasi optimal high performance blades for horizontal axis wind turbines". 18th International Conference of COMPLES, Milano, 1979.

Jesch, L. F. (1980) and R. T. Crossland: "Dynamic analysis of vertical axis wind turbine blades". in **Tagungsbericht des 3, Internationalen Sonnenforums – Freie Vorträge,** Hamburg 1980, DGS-Sonnenenergie Verlag, 450-457.

Jesch, L. F. (1980): "Linearised blade design for horizontal axis wind turbines". in **Tagungsbericht des 3, Internationalen Sonnenforums – Freie Vorträge,** Hamburg 1980, DGS-Sonnenenergie Verlag, 491.

Jesch, L. F. (1980) and D. Walton: "Reynolds number effects on the aerodynamic performance of a vertical axis wind turbine". Paper presented at the **3rd Int. Symp. on Wind Energy Systems** held in Copenhagen, Aug. 1980. BHRA Fluid Engineering, 323-332.

Jesch, L. F. (1981) and D. Walton: "Cyclic torque variation analysis in Darrieus rotors". Paper presented at the **International Colloquium on Wind Energy,** Brighton 1981 British Wind Energy Association.

Jesch, L. F. (1981), D. Mihas and J. Ramanathan: "Tower shadow effects of horizontal axis wind turbines in wind tunnel tests." In press.

Justus, C. G. (1978): "Methods for estimating wind speed frequency distributions". **Journal of Applied Meteorology, 17,** 1978, 350-353.

Kalke, S. (1977): "Die Tvindkraftmuhle". **Energie von Wind, Tagungsbericht,** Bremen 1977, Deutsche Gesellschaft für Sonnenenergie e.V., 323-327.

Kharitonov, V. P. (1981): "Problems of today pertaining to wind energy utilization". **Windirections,** 6, May 1981, 4-5.

Kreith, F. (1978) and J. F. Kreider: **Principles of Solar Engineering,** McGraw-Hill 1978, 778 p.

Lindley, D. (1980 A), P. B. Simpson, U. Hassan and D. T. Milborrow: "Assessment of offshore siting of wind turbine generators". Papers presented at the **3rd Int. Symp. on Wind Energy Systems** held in Copenhagen, Aug. 1980. BHRA Fluid Engineering, 17-42.

Lindley, D. (1980 B): "The adolescence of twentieth century wind energy system technology". **Journal of Industrial Aerodynamics, 5,** (1980), 207-221.

Lindley. D. (1981) and W. Stevenson: "The horizontal axis wind turbine project on Orkney" **Proceedings of the Third BWEA Wind Energy Conference.** Cranfield, April 1981, BHRA Fluid Engineering, 16-32.

Lipman, N. (1980), E. Bossanyi, P. D. Dunn, P. J. Musgrove, G. E. Whittle and C. Mclean: "Wind power integration studies, an overview of the Reading University programme". **Proc. of the 2nd BWEA Wind Energy Workshop,** Apr. 1980, Multi Science Publ. Co., 151-162.

Lissaman, P. B. S. (1980), T. G. Zambrano and S. N. Walker: "Wind energy assessment of the Palm Springs-Whitewater region, California, U.S.A.". Papers presented at the **3rd Int. Symp. on Wind Energy Systems** held in Copenhagen Aug. 1980. BHRA Fluid Engineering, 91-106.

Littler, J. G. F. (1979) and R. B. Thomas: "Wind power for domestic use in the United Kingdom." IEE Conf. Publ. 171, **Future Energy Concepts,** 1979, 313-316.

Ljugström, O. (1980): "L-180 Poseidon – a new system concept in vertical axis wind turbine technology". Papers presented at the **3rd Int. Symp. on Wind Energy Systems** held in Copenhagen Aug. 1980. BHRA Fluid Engineering, 333-355.

McVeigh, J. C. (1977): **Sun Power.** Pergamon Press 1977, 208 p.

Machens, U. (1980): "Beitrag der Windenergie zur Energiebilanz einer kombinierten Solar-Windkraft-Anlage" in **Statusbericht Windenergie,** VDI-Verlag, 1980, 131-145.

Meggle, R. (1980) and B. Schöbe: "Programmübersicht GROWIAN II" in **Statusbericht Windenergie,** VDI-Verlag 1980, 77-96.

Moore, D. J. (1979): "Offshore wind data". **Proc. of the 1st BWEA Wind Energy Workshop,** Apr. 1979, Multi Science Pub. Co., 199-207.

Musgrove, P. J. (1976): "The variable geometry vertical axis windmill". Proc. BHRA Wind Energy Symposium, Cambridge 1976.

Musgrove, P. J. (1978) and I. D. Mays: "Development of the variable geometry vertical axis windmill". Papers presented at the 2nd International Symposium on **Wind Energy Systems,** held in Amsterdam, October 1978. BHRA Fluid Engineering, Vol. 1, E4-39 – E4-60.

Nickols, W. R. (1979): "The design and construction of the Aldborough aerogenerator". **Future Energy Concepts,** IEE Conf. Publ. 171, 306-308.

Pedersen, B. M. (1980 and P. Nielsen: "Description of the two Danish 630 kW wind turbines, NIBE-A and NIBE-B, and some preliminary test results". Papers presented at the **3rd Int. Symp. on Wind Energy Systems** held in Copenhagen Aug. 1980. BHRA Fluid Engineering, 223-238.

Philbrick, D. (1979) and R. Morgan: "Wind anemometer Loan Program". Proc. ISES Silver Jubilee Congress in Atlanta 1979, Vol. 3, 2287-2290.

Putnam, P. C. (1948): **Power from the Wind.** Van Nostrand, New York, 1948.

Railly, J. W. (1981): "Prediction of the influence of mutual interference on the power from a wind turbine". Paper presented at the **International Colloquium on Wind Energy,** Brighton 1981, British Wind Energy Association.

Read, S. (1980) and D. J. Sharpe: "An extended multiple streamtube theory for vertical axis wind turbines". **Proc. of the 2nd BWEA Wind Energy Workshop,** Apr. 1980, Multi Science Pub. Co., 65-72.

Scholz, H. J. (1980) and K. Simhan: "Anmerkungen zum Steurungs – und Überwachungssystem von Windenergiekonvertern mit vertikaler Achse" in **Tagungsbericht des 3, Internationalen Sonnenforums – Freie Vorträge,** Hamburg 1980, DGS Sonnenenergie Verlag, 468-480.

Sforza, P. M. (1977): "Vortex augmentation of wind energy". **Wind Engineering, 1,** 1977, 186-197.

Sharpe, D. J. (1976): "An aerodynamic performance theory for the Darrieus wind turbine". **Int. Symposium on Wind Energy Systems,** Cambridge, 1976.

Sharpe, D. J. (1978): "Experiments with a Darrieus turbine". **Wind Power in the United Kingdom.** Proc. of a One Day Symposium, 1978. Multi Science Pub. Co., 8-20.

Spittler, W. (1980): "Entwicklung, Herstellung und Erprobung eines 52 m O/265 kW Windenergie-konverters" in **Statusbericht Windenergie** VDI-Verlag 1980, 47-51.

Swift-Hook, D. T. (1979 A): "Describing wind data." **Wind Engineering 3,** 1979, 167-186.

Swift-Hook, D. T. (1979 B): "A collaborative programme of field measurements on wind turbines". **Proc. of the 1st BWEA Wind Energy Workshop,** Apr. 1979, Multi Science Pub. Co., 5-12.

Swift-Hook, D. T. (1979) and R. H. Taylor: "Directional arrays of windmills". **Future Energy Concepts,** IEE Conf. Publ. 171, 1979, 290-293.

Templin, R. J. (1974): **Aerodynamic Performance Theory for the NRC Vertical Axis Wind Turbine,** NRC of Canada TR LTR-LA-160, 1974.

Thiele, H. M. (1980): "Fertigungsentwicklung, Bau und Test des GROWIAN I-Rotorblattes". in **Statusbericht Windenergie,** VDI-Verlag, 1980, 15-26.

Watson, G. R. (1979): "The self-starting capabilities of low-solidity fixed pitch Darrieus rotors". **Proc. of the 1st BWEA Wind Energy Workshop,** Apr. 1979, Multi Science Pub. Co., 32-39.

Watson, G. R. (1979) and G. R. Bainbridge: "The design construction & proving of a low cost 5 kW wind powered turbine for isolated applications". **Future Energy Concepts,** IEE Conf. Pub. No. 171, 1979, 277-281.

Willmer, A. C. (1979): "Low Reynolds number tests on the NACA 0015 section". **Proc. of the 1st BWEA Wind Energy Workshop,** Apr. 1979, Multi Science Pub. Co., 109-116.

Zelck, G. (1980): "Erstellung baureifer Unterlagen für einen 60 kW Vertikalachsen-Schnelläufer-rotor mit geraden Rotorblättern" in **Statusbericht Windenergie,** VDI-Verlag 1980, 273-288.

Chapter 8

APPLICATIONS

The concepts, the practice and the problems of solar and wind energy conversion were explained in the preceding pages. To illustrate specific points examples from all over the world were used.

On the following pages the British examples are shown. Some are included because of historic significance, others because of unusual design. There was no attempt made to give a comprehensive coverage of solar energy applications in Britain, as it is practised today. There are seven passive, three hybrid, twenty one active, two testing applications, three wind energy installations and one photo-voltaic application. Some of the applications which could not be included would further corroborate the message of this chapter: solar energy today is actively and skilfully pursued in Britain, photovoltaics and wind energy are becoming important areas for research and development.

The sequence in which the applications are presented is arbitrary, although within each group there is an approximate chronology of design concept. Information contained in the descriptive parts or in the technical specifications is presented as it was received and no attempt was made to verify details. The inclusion of an application does not imply approval of any sort, nor is there any value judgement about those applications which were not included. Wherever further information, in the form of easily available publications, was known, it was listed together with some points of contact: usually the designers and the operators of the installation.

8.1 **ST. GEORGE'S SCHOOL, WALLASEY**

One of the earliest examples in the U.K. of conscious passive design is the St. George's School at Wallasey, Cheshire. The original school was constructed in 1955 and in 1961 the building was extended. This extension, which houses classrooms and the gymnasium, was designed to maintain a comfortable thermal environment using heat gains from occupants, the lighting system and a large south facing double-glazed wall. The solar heated extension consists of a west wing, 230ft x 40ft x 23ft high which contains all the classrooms and the assembly hall, and an east wing 105ft x 50ft x 20ft high which contains the gymnasium and workshops. The high standard of thermal insulation on the northern walls and on the roof with the massive concrete construction give the building a high time constant. Although a detailed monitoring of the school has never been carried out, comparison with other schools of similar size and occupancy show significant reductions in the heating costs.

Designer: E. Morgan.

Further information: Hitchin, E. R. (1966), K. Thompson and C. B. Wilson: "St. George's County Secondary School, Wallasey". **J.I.A.V.E. 33,** Jan. 1966, 325.

Davies, M. G. (1975): "Heating buildings by winter sunshine". Conference on **Solar Energy Architecture and Planning,** April 1975, UK-ISES.

8.2 'DELTA' HOUSE, CHARSFIELD, SUFFOLK

Built in 1974, this house incorporates high standards of insulation, passive gain through south-facing glazing and high thermal mass. Thermal storage is provided by a heavy ground floor raft with a black tiled finish. The warm air from a 30 m^2 conservatory can be drawn into the house and the heat may be retained within the living portion of the house by the use of heavy curtains. Back-up heating is provided by off-peak electric storage heaters. It is claimed that over a third of the total radiation falling on the glazing can be collected as useful heat for the interior.

Architects: Cedric Green.

Monitoring: Cedric and Margaret Green.

8.3 MIDGLEY, YORKSHIRE

This is an interesting example of a builder wishing to experiment with some basic solar architectural principles when designing a house. The single storey house was designed by the builder and sold on the open market. The house is well insulated and has a high thermal mass concrete and stone structure, contains both direct gain and solar wall elements. The living room shown on the left of the picture (taken during construction) has a large south facing window, and the floor of the room is concrete covered with tiles. A solar wall is incorporated into both the right hand section of the south facing wall and in the west wall. Insulating shutters are used on the solar walls. The shutters are designed to track the sun allowing the maximum radiation through whilst closing up at night or in overcast conditions to retain the heat.

Designer: L. Neen.

8.4 ACORN CLOSE, BEBINGTON

This development owned by Merseyside Improved Houses consists of 14 houses and is situated at Higher Bebington on the Wirral. None of the houses are insulated to a high level, e.g. double glazing, draught proofing, insulation in floor, walls and ceiling. In addition they have a double glazed brick solar wall on the southerly aspect. The remaining five houses are of the same plan form as the solar houses but do not have a solar wall and were originally insulated to 1976 building regulations standards. In 1980 the insulation of the non-solar houses was increased, to correspond more closely to that of the solar house, by the addition of blown glass-fibre in the cavity. The distribution of the houses on the site is shown in figure 1. Five of the solar assisted houses are approximately south facing as is the terrace of non-solar houses. The other four solar houses consist of two semi-detached pairs facing respectively south-east and south-west. The houses are occupied by people in the age range of 60 years and over. Some 250 sensors are built into the houses, largely consisting of platinum resistance thermometers to measure structural and interior temperatures. A weather station is situated on one pair of houses. The sensors are connected by underground cable to a data logger on site which in turn is linked by a telephone line to a computer at the laboratory 25 miles away. Three separate electricity meters are provided for each house to monitor space heating and water heating in addition to total energy usage. Energy consumption is recorded weekly.

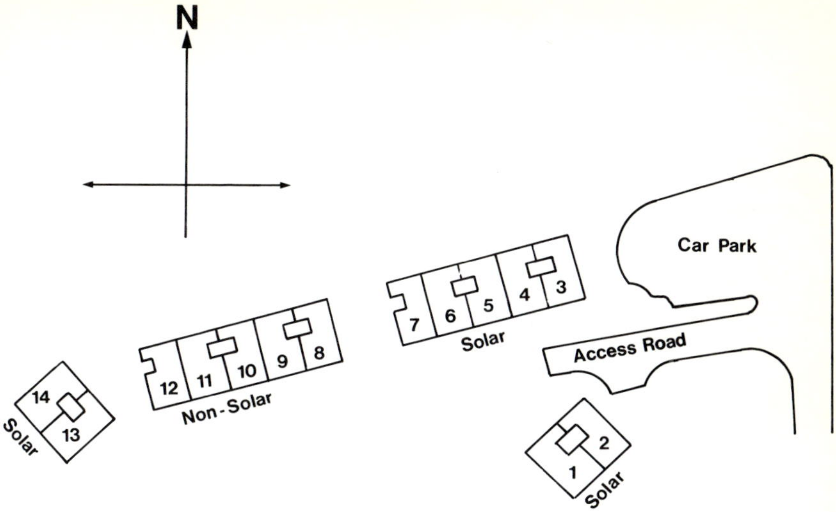

Acorn Close, Bebington, SITE PLAN.

Designers: Mr. P. G. Greenwood and Dr. H. S. Ward.

Monitoring: Pilkington Brothers Ltd., R & D Laboratories.

Further information: Greenwood, P. (1979) and H. Ward: "Solar houses for the elderly, Acorn House, Wirral. A development by Merseyside Improved Houses." in **The Passive Collection of Solar Energy in Buildings.** Conference (C19) London, UK-ISES 1979, 60-74.

Justin, B. (1980), A. G. Guy and G. Shaw: "Monitoring solar walls in occupied houses." in **Modelling & Monitoring Solar Thermal Systems.** Conference (C24) London, UK-ISES 1980, 94-104.

8.5 POULTON LANCELYN PRIMARY SCHOOL, WIRRAL

Partly as a result of the experiences gained from the solar heated St. George's School at Wallasey, the Wirral Metropolitan Borough Council has built this new primary school incorporating a number of low energy features. High standards of insulation have been built into the walls, ceiling and floor and the building has a high thermal mass to iron out temperature fluctuations. A heat recovery system based on the thermal wheel is also incorporated. The main solar features are specially designed splayed double glazed windows with internal shutters and a solar wall system. The solar wall is divided into six sections; four panels of 6 m x 2.5 m, and two panels of 4 m x 4 m. The structure of these walls consists of an outer double glazed skin situated 10 cm in front of a 200 mm thick outer brick wall which acts as the solar absorber. This is followed by a 50 mm air gap and an inner leaf of 100 mm engineering brick faced with 75 mm of polystyrene with a plasterboard finish. Warm air is drawn from the cavity in the brick wall to heat the building. A small gas fired boiler is used to provide domestic hot water and back-up heating through a heater battery in the incoming air duct.

Architects: Metropolitan Borough of Wirral, Department of Development, Architectural Division.

Monitoring: Metropolitan Borough of Wirral, Department of Development, Energy Conservation Unit in conjunction with Liverpool Polytechnic.

8.6 LINFORD SUMMERHAYES, MILTON KEYNES

This is a development of eight detached 4 bedroom houses which has been constructed for sale on the open market. The houses have 100 mm of fibreglass in the cavity, 150 mm in the roof and insulation in the floor. Double glazing has been used throughout the house. The solar heating has been achieved by increasing the glazing area on the south side of the building, while glazing on the north side being reduced. Additional heating is obtained by a high-efficiency gas boiler and a wet radiator system.

The houses are monitored in depth. The energy consumptions of the space heating and hot water are recorded. Temperature sensor and heat flow meters were incorporated into the structure of the houses during the construction phase and details of room temperatures and of structures will be recorded.

Architects: Energy Conscious Design.

Monitoring: Open University, Energy Research Group.

Further information: Doggart, J. (1980) and R. Everett: ''Passive solar housing at Milton Keynes.'' in **European Community Solar Housing.** Conference (C23) London UK-ISES 1980, 40.

8.7 PENNYLANDS, MILTON KEYNES

The Pennylands project represents the largest monitored passive space heating project in Europe. It consists of 177 local authority houses set out as shown in the diagram. The effect of orientation is one of the factors being studied. As it can be seen from the diagram a conscious attempt has been made to orient the house in a southerly direction. However, the estate does not give an atmosphere of being over structured and the skilful use of landscaping has produced a pleasant atmosphere. Apart from orientation the group is divided into 94 houses with high insulation levels (2″ cavity wall insulation, single glazing, 3″ roof insulation) and 83 houses with very high insulation levels (4″ cavity wall insulation, double glazing, 6″ roof insulation – some also equipped with insulating window shutters). A further factor being studied is the influence of window size and the distribution of window area, e.g. concentration of glazing on the south side. All the houses are being monitored for their energy consumption. In addition a number of houses are being equipped with electronic thermometers. These will be read weekly along with the gas and electricity meters. The conventional heating system consists of a high-efficiency low thermal capacity gas boiler and water filled radiators. A statistical analysis will be carried out on the results in an attempt to separate the effects of the various insulation and direct solar gain factors.

Architects: Milton Keynes Development Corporation.

Monitoring: Open University, Energy Research Group.

Further information: 'Passive Solar in Milton Keynes', R. Everett, Open University.

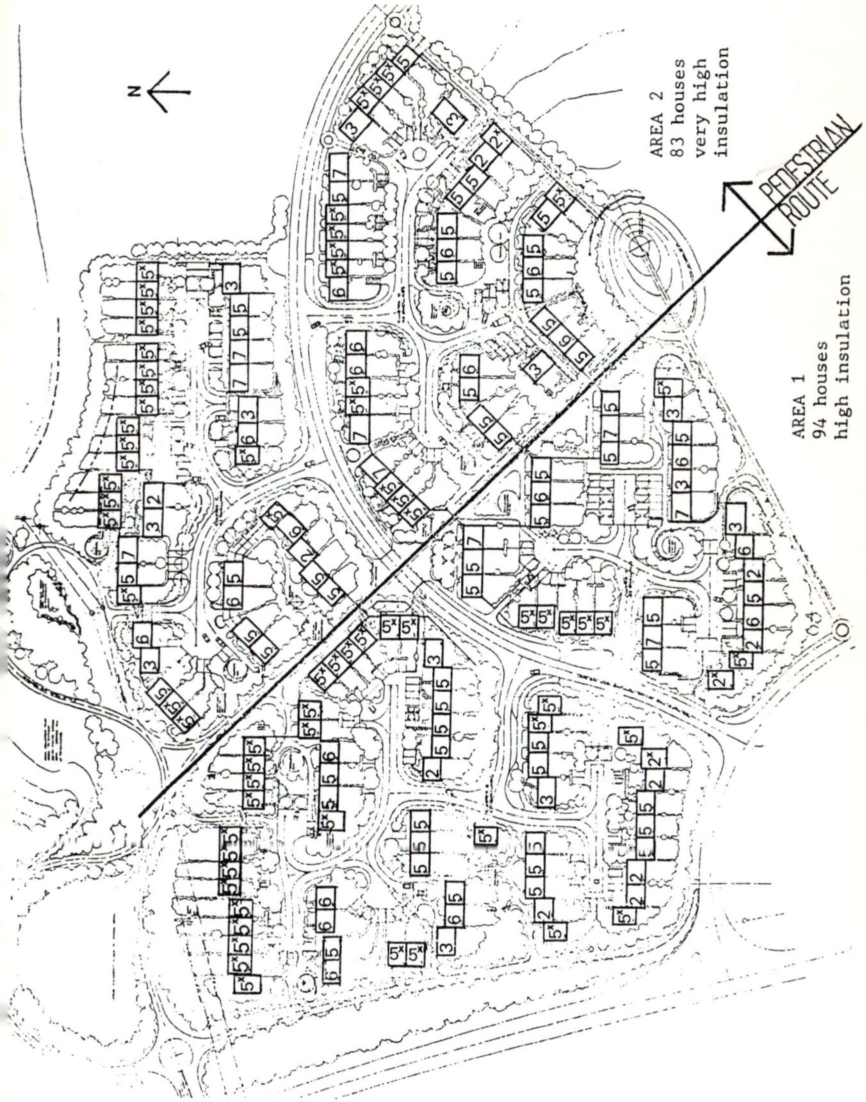
N
AREA 2
83 houses
very high
insulation
PEDESTRIAN
ROUTE
AREA 1
94 houses
high insulation

8.8 FUTURE HOME 2000, MILTON KEYNES

This four-bedroom detached timber frame house was featured in a series of television programmes by the BBC 2's 'Money Programme' in 1981. The building is insulated to approximately twice the standard required by building regulations. All the windows on the exposed outside walls are double glazed and incorporate an infra-red reflecting glass. One of the special solar features of the house is the double height conservatory which is built onto the south side. Warm air from the conservatory can be fed into the house by the use of small fans. It has been estimated that up to 2000 kWh of energy can be supplied from the conservatory. The back-up heating is provided by a conventional water filled radiator which can be fed from a high-efficiency gas boiler, or a solid fuel fire, or a total Energy Module (TOTEM) which consists of a Fiat 127 car engine adapted to run on either natural or liquid gas. The roof of the conservatory also contains two banks of selectively coated solar collectors which contribute towards the hot water requirements. The collectors are not of the conventional flat plate design but contain an integral water store. This eliminates the need for a circulating pump and a control unit.

Architects: Energy Conscious Design.

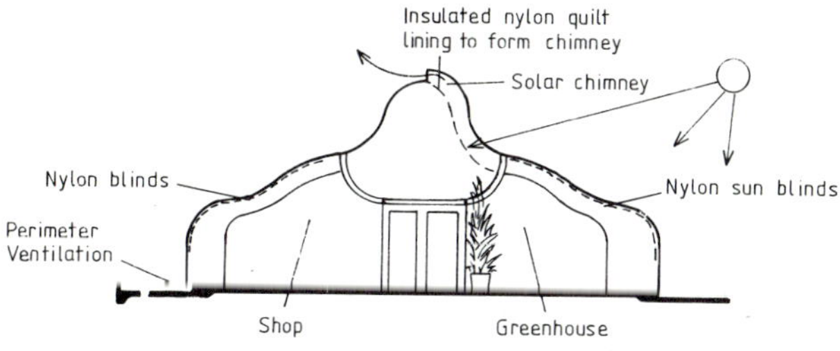

8.9 COLONNADES GARDEN CENTRE, LONDON

This £30,000 greenhouse was specially designed to make efficient use of solar radiation for heating and cooling. The building with a total floor area of 140 m^2 contains both a shop and the greenhouse area, see diagram. It is glazed in twin-wall polycarbonate sheeting and is equipped with 55 m insulating blinds to regulate heat flows through the fabric. Ventilation is induced by a chimney built into one half of the roof, and air changes of about 15 ac/hr have been measured.

Architects: Terry Farrell Partnership.

8.10 IDEAL HOME SOLAR HOUSE, MILTON KEYNES

This house contains both passive and active solar elements. The house, which consists of 150 m^2 floor space excluding the south facing two-storey conservatory is insulated to a U-value of 0.35 $W/m^2.°C$. Most of the windows are triple glazed. Overheating of the conservatory in summer is prevented by the use of a shading device and the use of high thermal mass in the floor and walls. All the main rooms are situated on the south side of the house whilst the services and ancilliary rooms are situated on the north side. Apart from the passive gains the house is also equipped with 27 m^2 of selectively coated flat plate collectors which are coupled to a latent heat storage system as shown in the diagram.

Architects: Dominic Michaelis Associates.

Further information: Copper Development Association, Information Sheet No. 3, May 1981.

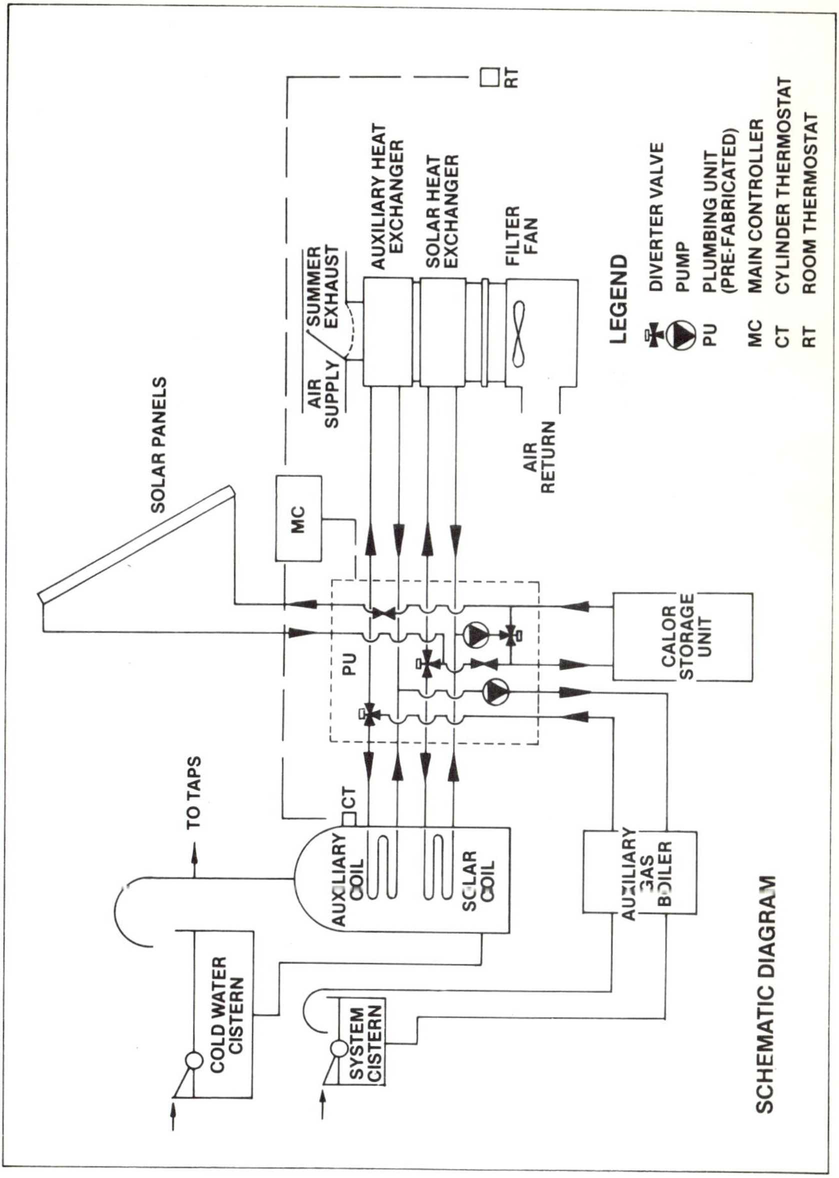

SCHEMATIC DIAGRAM

8.11 **BRADVILLE, MILTON KEYNES**

This house was probably the first active solar space heating system to be built and monitored in the U.K. A standard two storey terraced house was converted into a solar experimental house by the addition of 40 m^2 of Rollbond aluminium collectors into the south facing mono-pitch roof. The collectors are coupled to two 2.1 m^3 storage tanks via a direct circulating system. Space heating is provided by fan convectors operating either directly from the solar store or boosted by a gas boiler. Preheating of the domestic hot water is achieved by passing the cold water feed through heat exchangers in the solar storage tanks before entering the hot water cylinder. The system has been monitored and modifications carried out included painting the absorbers matt black to improve the absorbtivity, the introduction of a non-return valve to prevent reverse circulation through the collectors and an improved control strategy for the heating system. The overall system performance has progressively improved from 31% in 1975/76 to 45% 1978/79.

Architects: Milton Keynes Development Corporation and the Polytechnic of Central London.

Monitoring: Polytechnic of Central London.

Further information: Hodges, D. (1976): "Analysis of the Performance of the Solar House at Milton Keynes." Conference (C8), UK-ISES, April 1976.

Horton, A. (1979): "Improved performance of the Milton Keynes solar house." **Sun at Work in Britain, No. 9,** 1979, 14-23.

MKDC/PCL SOLAR HOUSE

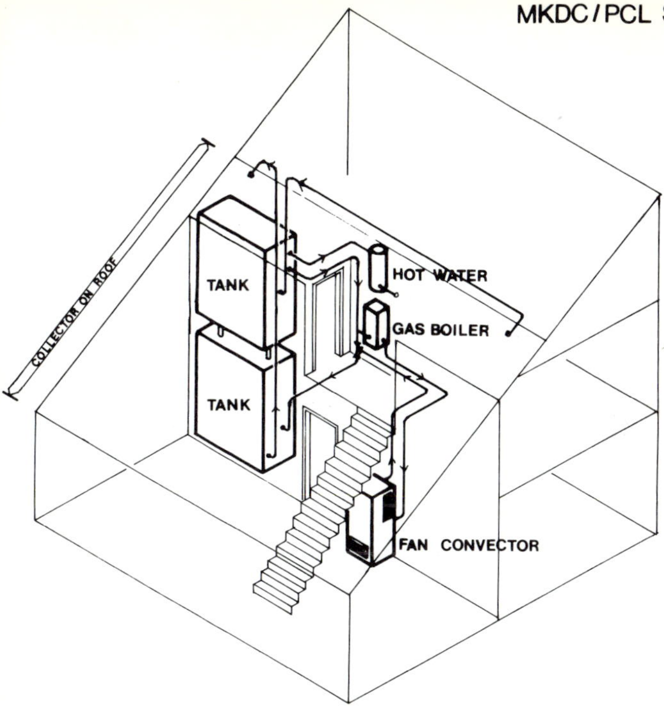

POSITION OF MAJOR COMPONENTS

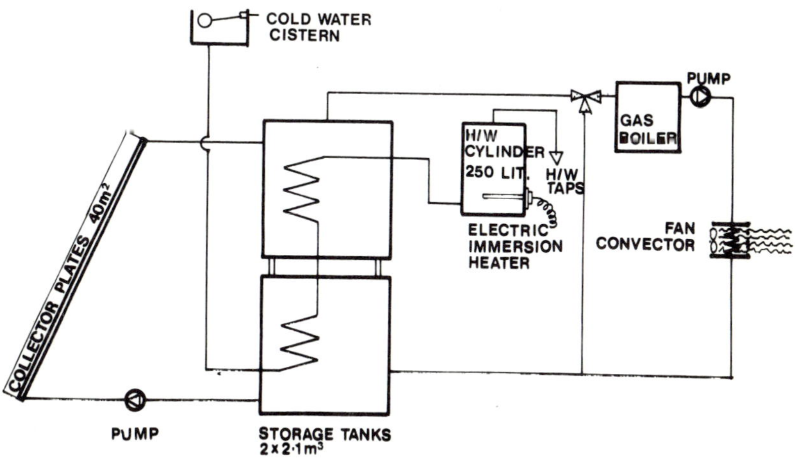

SYSTEM DIAGRAM

8.12 WIMPEY SOLAR HOUSE, HAYES

Although there are many examples of active spacing heating systems in the U.K. which use aqueous or oil based heat transfer fluid there are few examples of air based systems. The experimental solar house at the Wimpey Laboratories has 25 m^2 of air collector in the south facing roof. The collector is coupled to a No-Fines concrete heat store and space heating is achieved by ducted warm air. The house has been insulated to a high level so that it will require approximately one half of the heating it would need if constructed to 1976 Building Regulations. In addition to space heating the house is also equipped with 6 m^2 of conventional water heating panels incorporated into the roof just above the eaves. The House is being monitored and occupancy is being simulated using heaters controlled by a micro-computer.

Architects: Wimpey Laboratories in cooperation with Wimpey Group Services.

Monitoring: Wimpey Laboratories.

8.13 **SOLAR HOUSE, BLYTH**

Completed in 1976, this house was built speculatively and sold on the open market. A standard 150 m^2 four-bedroom detached house was modified by the addition of extra insulation and by the incorporation of 40 m^2 of solar collectors into the 22.5° pitched roof. A 4 m^3 water store is situated below ground in the back garden. The primary solar circuit is a pressurised antifreeze system. Heat from the solar store is extracted from the store via a heat exchange to a series of forced air convector heaters. Back-up heating is obtained from an oil fired boiler.

Architects: Design and Materials Ltd.

Further information: Copper Development Association, Information Sheet No. 2, Nov. 1977.

8.14 FORESTDALE, CROYDON

Built in 1977 these three houses were some of the first ones in the private housing sector to be fitted with solar water heaters. Each house has a total of 8 m^2 of copper absorbers integrated into the roof. The heat is transferred from the collectors to a 230 litre storage tank by an indirect pressurised circuit containing an anti-freeze mixture. An anti-boiling safety device is built into the primary circuit. Usable heat produced per annum is estimated to be around 2500 kWh.

Architects: Masini Franklin Partnership.

Further information: Copper Development Association, Information Sheet No. 1, November 1977.

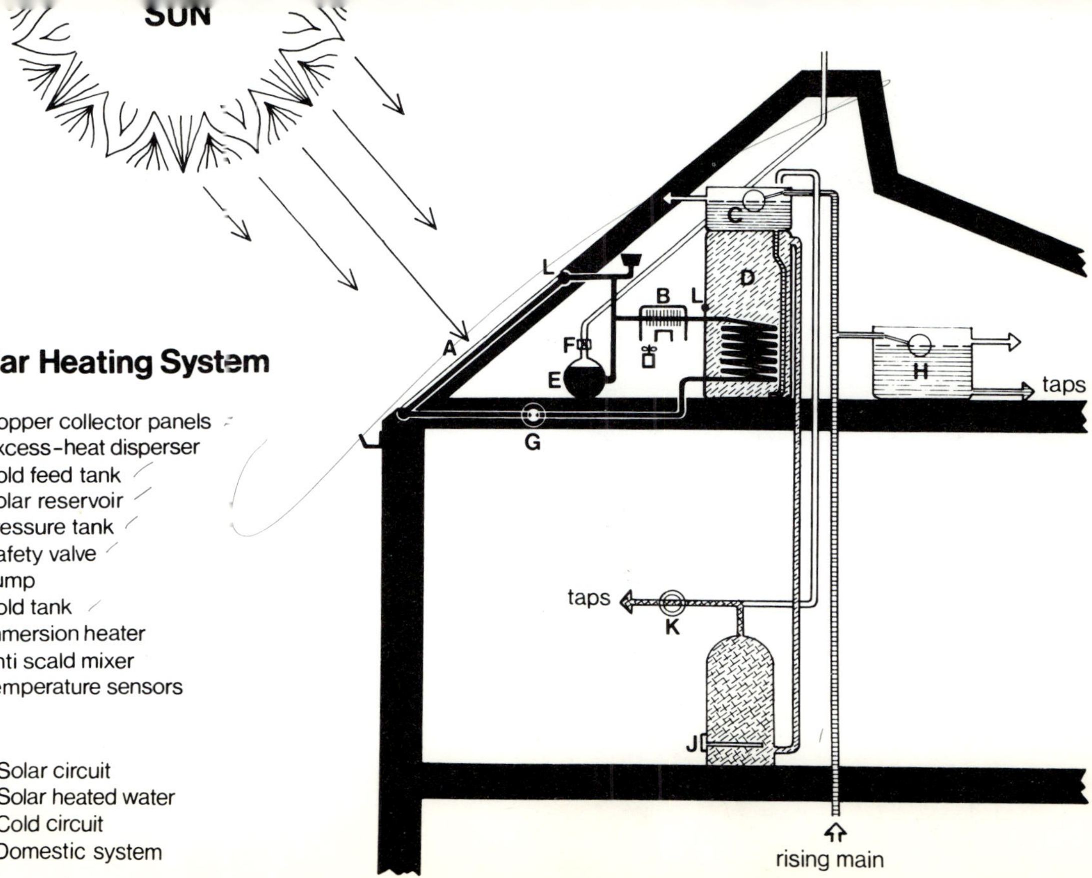

Solar Heating System

A Copper collector panels
B Excess-heat disperser
C Cold feed tank
D Solar reservoir
E Pressure tank
F Safety valve
G Pump
H Cold tank
J Immersion heater
K Anti scald mixer
L Temperature sensors

Solar circuit
Solar heated water
Cold circuit
Domestic system

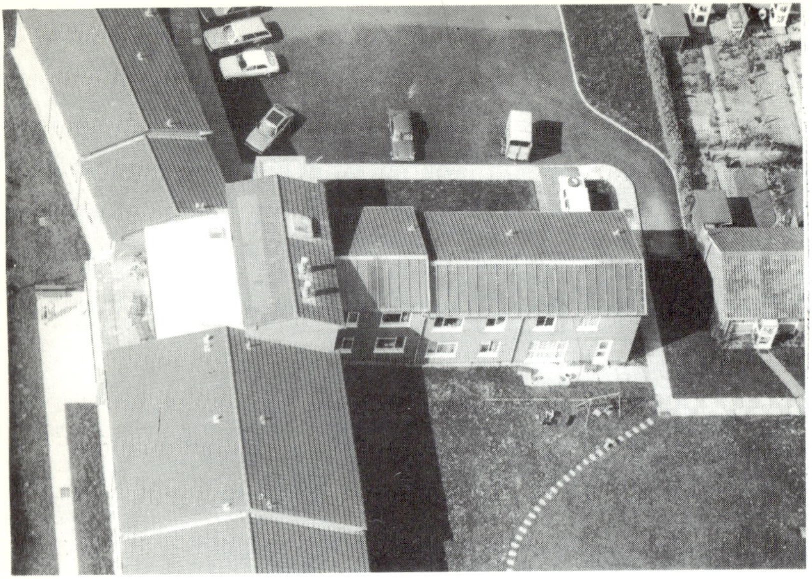

8.15 **ROWHEATH, BIRMINGHAM**

Owned by the Bournville Village Trust this sheltered accommodation for elderly people was opened in 1977. It consists of 40 single bedroom flats, communal day rooms, guest rooms and a warden's flat. An area of 40 m^2 rollbond aluminium collectors under patent glazing forms part of the roof structure. The collectors are coupled to two 1.5 m^3 hot water cylinders by a closed circuit filled with anti-freeze solution. The system provides pre-heating for the water supply to the laundry room and the flats. The thermal performance of the system is being monitored and a meteorological station has been established on the roof.

Architects: Bournville Village Trust.

Designer: Calorsol Ltd.

Monitoring: Solar Energy Laboratory, University of Birmingham.

Further information: **Solar Energy Projects,** A Bournville Village Trust Publication, 1981.

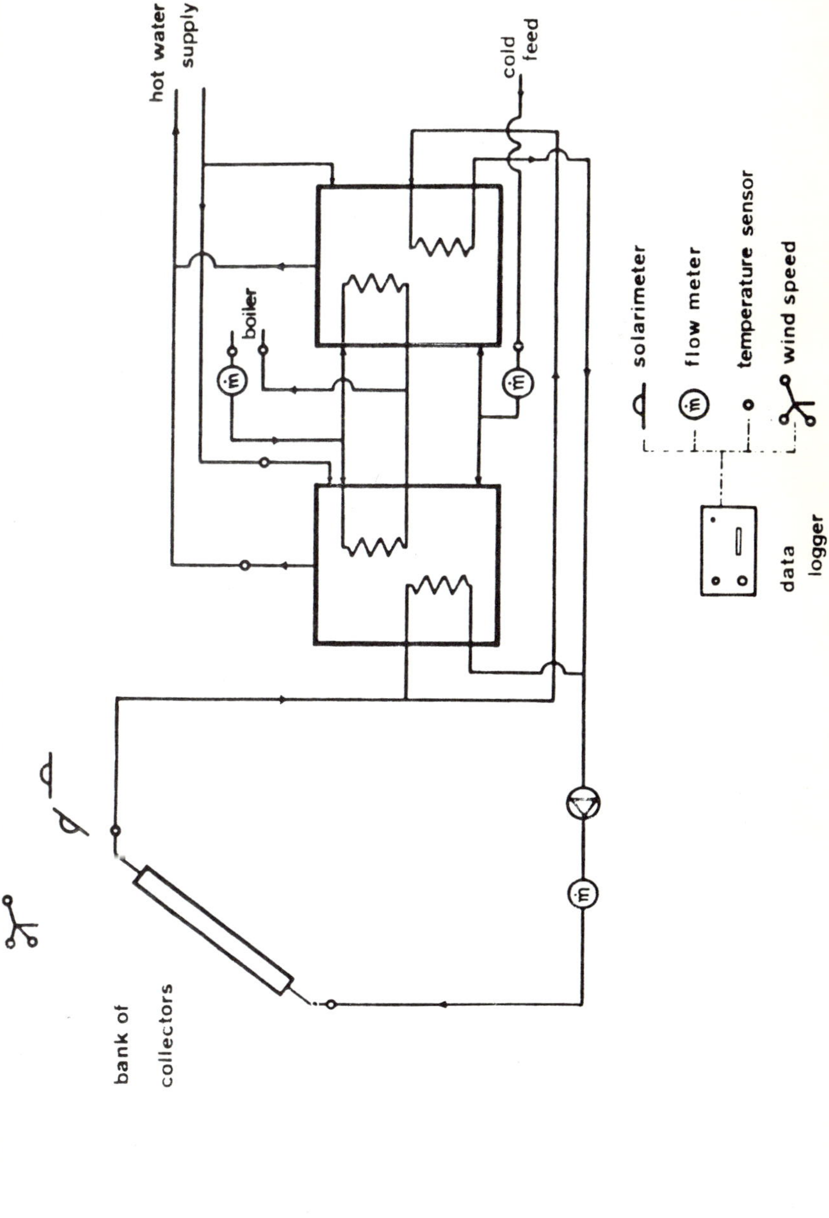

hot water
supply
cold
feed
boiler
bank of
collectors
solarimeter
flow meter
temperature sensor
wind speed
data
logger

8.16 STUDENTS FLATS, CARDIFF

A block of four student flats at University College, Cardiff, was fitted in 1977 with 25 m^2 of flat plate collector to supply hot water to showers, wash basins and kitchens for about 20 students. All copper collectors, half having selective coatings, and an 800 litre solar preheat tank are used in a direct system as shown in the diagram. When the temperature of the collectors falls to 3°C the system drains down automatically to prevent frost damage to the system. The system is fully instrumented and a small meteorological station is situated on the roof. A comparison of the energy savings is also made with the other blocks of flats in the same student hall. The monitored data are being used to validate computer models of the solar system and results so far have shown good agreement between predicted and measured performance. The solar contribution to the hot water heating load is about 30% when averaged over the year.

Monitoring: Solar Energy Unit, University College, Cardiff.

Further information: Taylor, P. J. (1979): "Monitoring solar water heating systems." **Helios** No. 6, July 1979.

Taylor, P. J. (1980): "Monitoring a large solar water heating system." in **Solar Energy in the 80s.** Conference (M2) Birmingham, UK-ISES Midlands Branch 1980, 79-84.

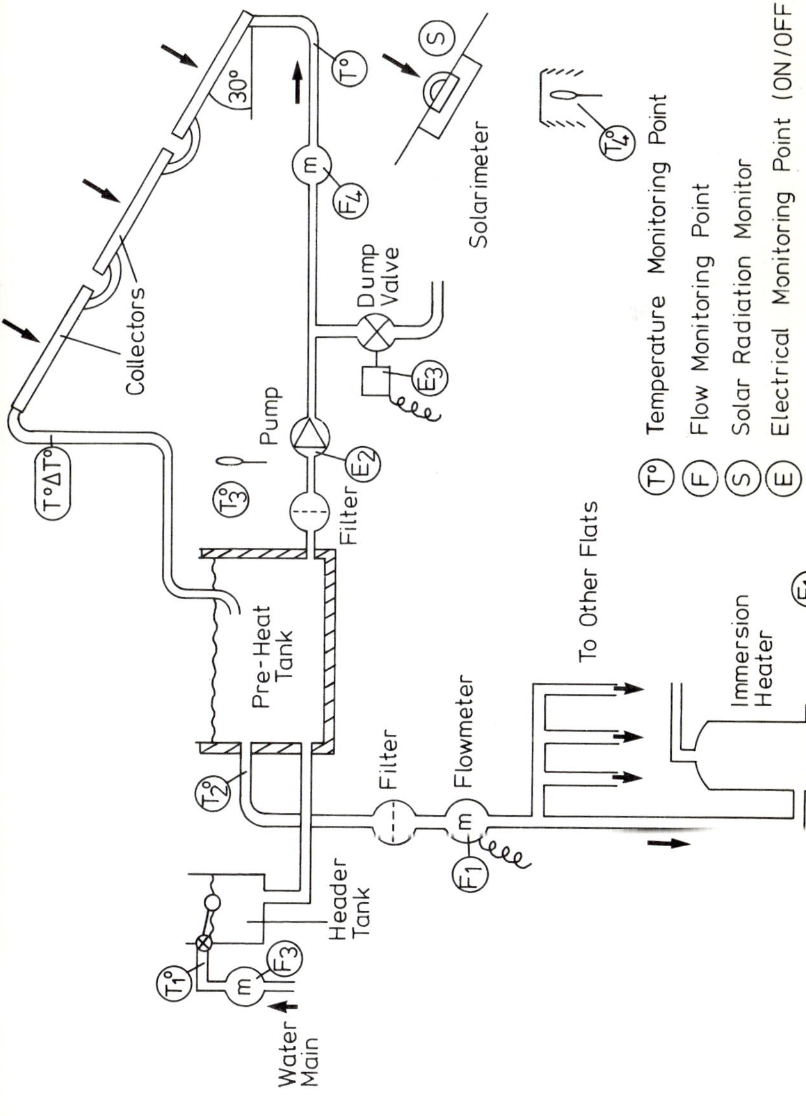
Collectors
30°
T°
T°ΔT°
S
Solarimeter
T_4^o
F_4
m
Dump Valve
E_3
Pump
E_2
T_3^o
Filter
Pre-Heat Tank
To Other Flats
E_1
Immersion Heater
Filter
Flowmeter
T_2^o
m
F_1
Header Tank
F_3
T_1^o
m
Water Main
T° Temperature Monitoring Point
F Flow Monitoring Point
S Solar Radiation Monitor
E Electrical Monitoring Point (ON/OFF)

8.17 **SOLAR HOUSE, MACCLESFIELD**

This house did much to stimulate interest in solar energy and other forms of energy technology. Often referred to as the 'Granada house' it featured in a series of television programmes in 1975, which showed the conversion of an old coach house to a two storey house. The main solar features are a 42 m^2 solar collector and an attached conservatory. The solar collectors form the roof and are of the single glazed open trickle type constructed from corrugated aluminium. The collector feeds three storage tanks, a 0.4 m^3 domestic hot water tank, a 0.7 m^3 high temperature tank and a 3 m^3 space heating tank. A 1.5 kW heat pump operating on off-peak electricity can be used to up-grade the heat from a 2 m^3 dump tank and transfer it to the space heating store. Space heating is provided by a wet radiator system heated by a solid fuel boiler. Other features of the house include high levels of insulation external to the solid walls and an air-to-air heat recovery unit. The performance of the system and its components has been monitored. Direct solar contribution to space heating has been measured at 2% with a further 10% solar contribution to the heat pump input. Solar contribution to the domestic hot water was found to be 12%.

Architect: D. R. Wilson.

Monitoring: The Electricity Council Research Centre, Capenhurst.

Further information: Wilson, D. R. (1977): "Solar collection and storage techniques in a house conversion at Macclesfield for Granada T.V." Conference (C13) London UK-ISES 1977.

Stephen, F. R. (1980): "Performance monitoring a low energy house, Macclesfield." **Ambient Energy 1,** January 1980, 29-46.

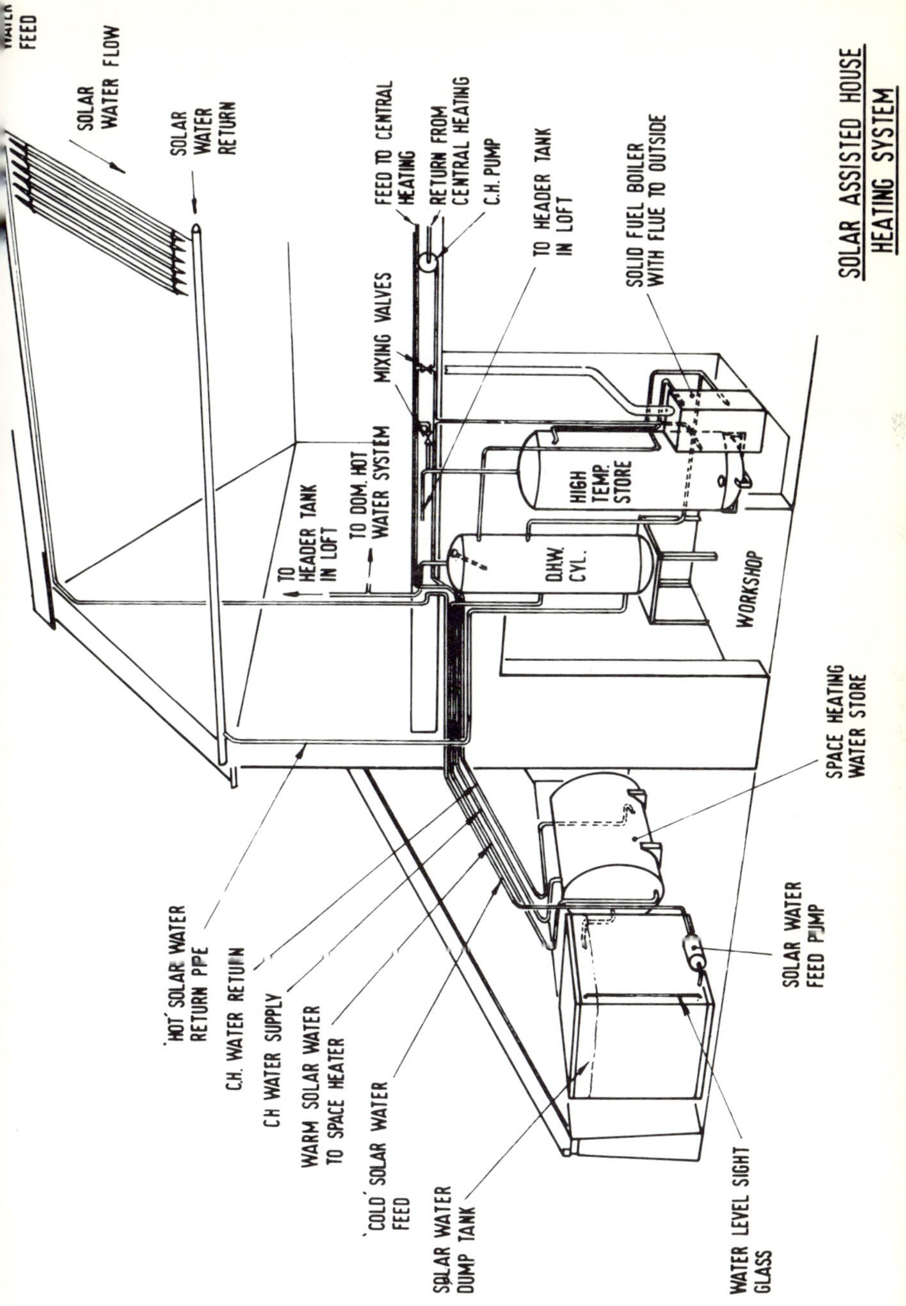
FEED
SOLAR WATER FLOW
SOLAR WATER RETURN
FEED TO CENTRAL HEATING
RETURN FROM CENTRAL HEATING
C.H. PUMP
TO HEADER TANK IN LOFT
SOLID FUEL BOILER WITH FLUE TO OUTSIDE
SOLAR ASSISTED HOUSE HEATING SYSTEM
MIXING VALVES
TO DOM. HOT WATER SYSTEM
TO HEADER TANK IN LOFT
HIGH TEMP. STORE
D.H.W. CYL.
WORKSHOP
SPACE HEATING WATER STORE
'HOT' SOLAR WATER RETURN PIPE
C.H. WATER RETURN
C.H WATER SUPPLY
WARM SOLAR WATER TO SPACE HEATER
'COLD' SOLAR WATER FEED
SOLAR WATER DUMP TANK
SOLAR WATER FEED PUMP
WATER LEVEL SIGHT GLASS

8.18 **WHATELEY ROAD, LONDON**

Of the several trial solar heating installations carried out in London, the scheme at Whateley Road is probably the most well known. Domestic hot water systems have been installed in 14 terrace houses. Twelve of these houses are fitted with 5 m^2 of single glazed matt black collectors coupled indirectly to a 440 litre hot water cylinder. A further house is equipped with the same basic system but uses 10 m^2 of absorber. The remaining house in the group of 14 has a system which is based on 5 m^2 of collector indirectly connected to a 160 litre preheat cylinder. The water from this cylinder can either feed the main hot water cylinder or can be drawn off directly through an independent series of 'solar taps' as shown in the diagram. The performance of all the systems is being monitored in detail.

Architects: South London Consortium, Energy Group.

Monitoring: South London Consortium, Energy Group.

Further information: Makkar, L. (1980): ''Monitoring of three tap and hybrid solar water heating systems at Whateley Road, London.'' in **European Community Solar Housing.** Conference (C23) London, UK-ISES, October 1980, 30-39.

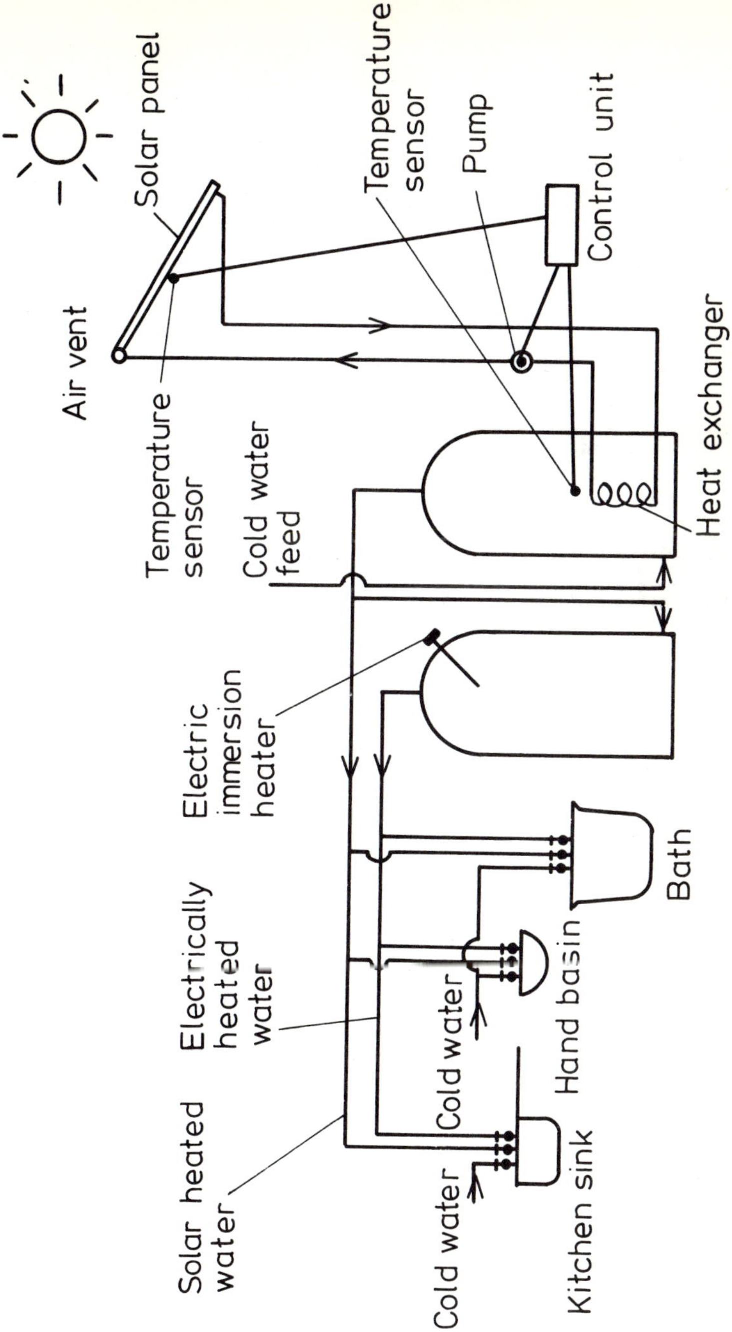

Solar panel
Temperature sensor
Pump
Control unit
Air vent
Temperature sensor
Heat exchanger
Cold water feed
Electric immersion heater
Bath
Electrically heated water
Hand basin
Cold water
Solar heated water
Cold water
Kitchen sink

8.19 ST. MARY AND ST. ANNE'S GIRLS' SCHOOL, ABBOTS BROMLEY

In 1978 it was decided to cover the 20 year old outdoor swimming pool. The opportunity was taken at the same time to utilise solar energy for heating. Three rows of 21 panels cover the roof giving a total area of 128 m^2 of collector area. The collectors were made in Israel and consist of aluminium fins mechanically bonded to parallel copper tubes. The pool has a surface area of 240 m^2.

Architects: T. H. Thorpe and Partners.

Designers: Electra Airconditioning Ltd.

8.20 DALE FORT FIELD CENTRE, DYFED

A recent addition to the hostel accommodation at the Dale Fort Field Centre, South Wales, afforded the opportunity to instal a solar water heating system. Approximately 23 m^2 of flat plate collectors are used to preheat a 1000 litre water storage vessel and an oil based heat transfer fluid is used in the indirect solar circuit. A heat meter is used to monitor the solar contribution to the hot water supply. The energy consumed by the circulating pump and the solar radiation falling on the collector bank is also recorded. For more detailed monitoring the system is also equipped with temperature sensors placed throughout the solar circuit.

Architects: Leonard Baart Associates.

Monitoring: Pilkington Brothers Limited, R & D Laboratories.

8.21 **STUDENTS HOSTEL, KIRKCALDY, FIFE**

The active solar system at Kirkcaldy Technical College hostel has a number of unusual features. The building is circular in plan and solar panels are arranged in two half quadrants, one facing SSE and one SSW. A multiple tank storage system enables a combination of collector array and storage volume to be achieved. The collectors are constructed from a super plastic aluminium alloy which is allowed to deform over a stainless steel serpentine. In total 40 panels each 1.8 m high by 1 m wide grouped in ten groups of four panels make up the 72 m^2 panel array. Frost protection is by drain-down of the panels although electric heating, warm water heating or anti-freeze can be used. The system has been instrumented and system performance and the effect of various control strategies is being monitored. The installation has been operating successfully for almost two years.

Architect: Fife Regional Council, Department of Architecture.

Monitoring: Mechanical Engineering Department, Napier College, Edinburgh.

Further information: Watson, A. (1979): ''Solar hot water for a Scottish hostel.'' in **Practical Applications of Solar Energy in Industry and Commerce.** Conference (M1) Birmingham, UK-ISES Midlands Branch, 1979.

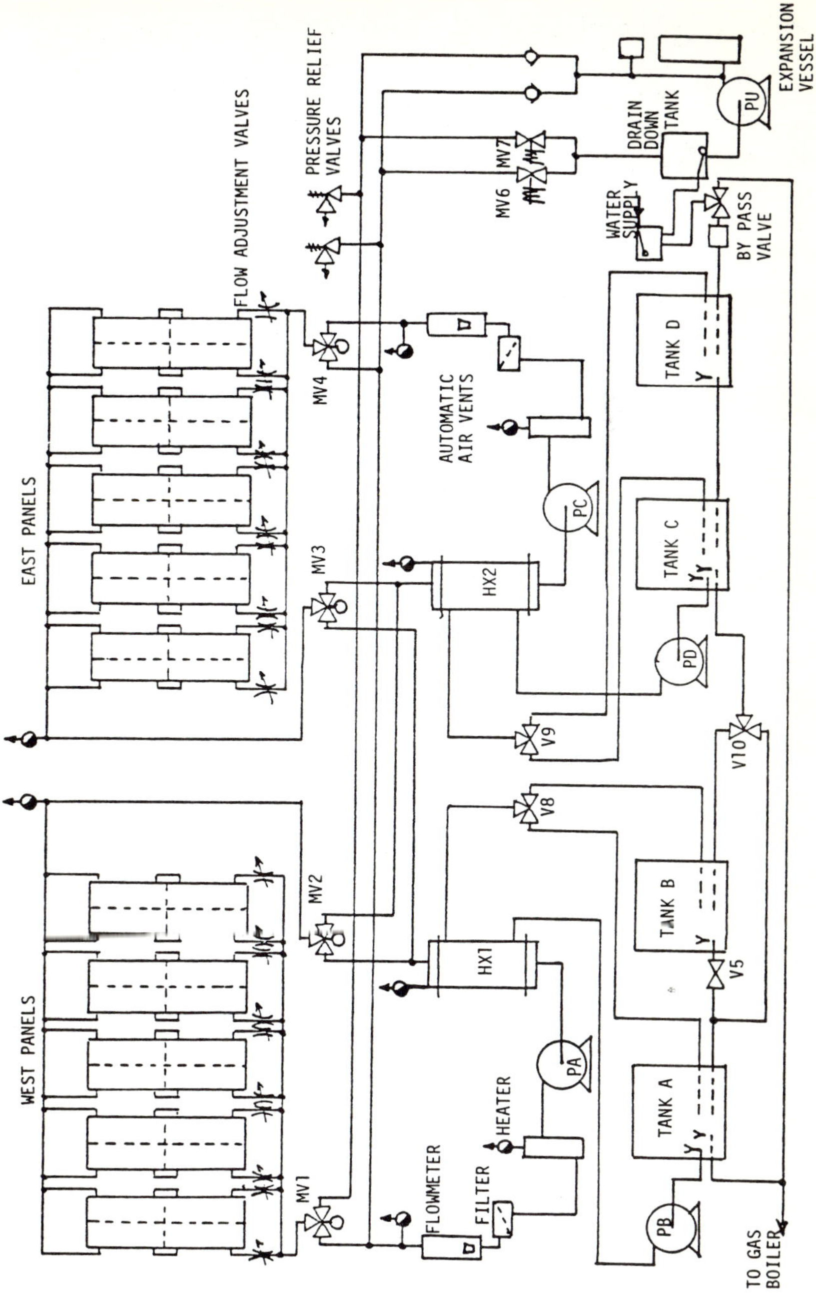

HYDRAULIC CIRCUIT LAYOUT

8.22 ROSEFIELDS, BIRMINGHAM

An installation of two banks of collectors, one facing south east and one facing south west, has been incorporated into this sheltered accommodation owned by the Bournville Village Trust. Opened in 1979, the building has 28 one person flats, 7 two person flats, and 8 bedsitting room flats. The two banks of collectors have their own pumps, heat exchangers and control devices but feed a common system of storage tanks. The total collector area is approximately 113 m^2. Energy from the primary circuits is transferred by means of heat exchangers to the secondary circuit which has three storage tanks, two 4.5 m^3 and one 1.5 m^3. These tanks may be used to bring about forced stratification. The system is controlled by a purpose built computer which interrogates 80 sensors and makes operational decisions according to preset strategies, or can change strategies in a heuristic, self adaptive manner. The roof-mounted meteorological station, which has four solarimeters, provides environmental data on solar radiation, cold water temperature, air temperature, humidity, wind velocity and direction. The installation is used to validate computer models of the solar heating system. Performance is being monitored and the information transmitted back to Birmingham University for analysis via a telephone line to the Computer Centre.

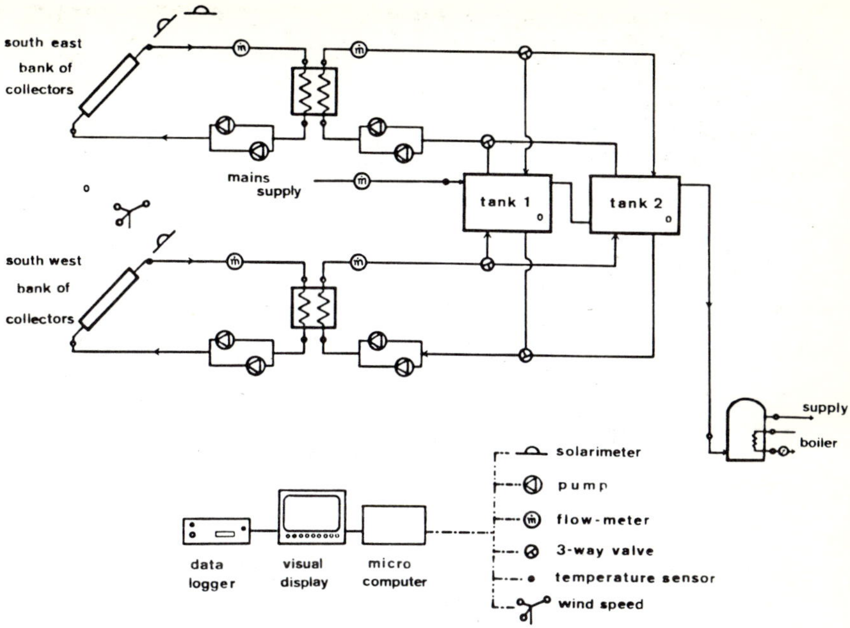

Architects: Bournville Village Trust.

Monitoring. University of Birmingham.

Further information: Jesch, L. F. (1979) and A. Zekios: "Adaptive control for a multiple tank, multiple temperature solar energy storage system." in **Sun II.** Proc. ISES Silver Jubilee Congress in Atlanta 1979, Pergamon Press, Vol. 1, 852-856.

Jesch, L. F. (1981) and T. W. Greeves: "Solar energy storage in forced stratification tanks." Papers presented at the **International Conference on Energy Storage,** Brighton 1981, BHRA Fluid Engineering, Vol. 1, 247-257.

Jesch, L. F. (1981), T. W. Greeves and D. Haralambopoulos: "Simulation of optimal control strategies for large solar water preheat systems." Proc. ISES Solar World Forum in Brighton, 1981.

8.23 PRIMARY SCHOOL, PORT ISAAC, CORNWALL

The 92 m^2 of flat plate collectors together with the water store are housed in a building which is separate from the school and the heated water is transferred to the school's hot water system via underground pipes. The system has been designed to preheat the domestic hot water, provide a part of the space heating and it is planned to heat a trainer swimming pool, when installed, during the summer months. The collectors are fixed in two rows, one at 35° to the horizontal and one at 65°. The collectors are coupled to a 11.3 m^3 water store. Auxiliary heating is provided by an oil fired boiler but when the solar heated water exceeds 40°C this is also pumped directly to the fan convectors and natural convectors that provide the space heating. The thermal performance of the system is being monitored.

Architects: Cornwall County Council, Department of Architecture.

Monitoring: University of Bath.

8.24 **GREAT LINFORD, MILTON KEYNES**

A total of nine houses make up this private development. All nine houses make use of passive gains through large south facing windows. In addition, however, some of the houses have active solar systems. The development is being used to study the differences in thermal performance between three groups of houses. One set of three (see above) contains 40 m^2 of solar collectors coupled to a latent heat storage system. The heat from the store is used to supplement the warm air heating system. A further set of three houses has 18 m^2 of solar collectors per house which assist a heat pump system. A large underground water store and a waste water store are coupled by the heat pump to a smaller day store. Space heating is provided by under floor heating and ducted warm air. The remaining three houses which are to the same plan are used as a control group for comparison. Detailed monitoring is being carried out on the systems and weather data are recorded on site.

Architects: Solar Energy Developments.

Monitoring: Calor Group Ltd., (active space heating with latent heat store). John Laing R & D Ltd., (solar assisted heat pump space heating).

Further information: Newton, G. (1980): "Monitoring of solar space heating and hot water systems at Milton Keynes." in **Modelling and Monitoring Solar Thermal Systems.** Conference (C24) London, UK-ISES, 1980.

8.25 NORTHLANDS ESTATE, BASILDON

As part of a programme to monitor the performance of solar systems, 12 houses on a new estate in Basildon have each been equipped with 5.4 m^2 of solar collectors. A water filled indirect drain-down circuit is used to transfer the energy collected to a 0.27 m^3 cylinder. In addition the houses also feature a communal heat pump which provides the space heating and the domestic hot water. The performance of the system is being monitored.

Architects: Ahrends, Burton and Koralek.

Monitoring: Building Research Establishment.

8.26 **ORTON SOUTHGATE, PETERBOROUGH**

As part of the Peterborough Development Corporation's programme of studying energy conservation techniques, in 1980 four new houses were fitted with flat plate solar collectors for domestic hot water pre-heating. Each house has 5.4 m^2 of selectively coated collectors which are coupled to a 0.2 m^3 hot water cylinder by a direct drain-down circuit. The circulation in the solar circuit is also stopped when the solar circuit temperature reaches 60°C thereby preventing scalding.

Architects: Peterborough Development Corporation.

Monitoring: Peterborough Development Corporation.

8.27 **BUCKSTONE, EDINBURGH**

Two semi-detached 3 bedroom houses have been designed to benefit from both active and passive solar gains. These houses, which were sold on the open market, incorporate 28.2 m^2 of single glazed copper tube and fin collectors to contribute to both the space heating and domestic hot water. An aqueous antifreeze solution is used in the solar primary circuit and the heat is stored in a 0.4 m^3 phase change heat store which utilises stabilised Glauber's salt. Auxiliary heating is supplied by a gas-fired ducted warm air heating system.

Architects: Solar Energy Developments.

Monitoring: Calor Alternative Energy.

8.28 ALOYNE HOUSE HOSTEL BLOCK, ABERDEEN

This 15 person hostel block owned by the Camphill Village Trust was completed in April 1980. The building, which is two storey and has ten bedrooms, has high levels of insulation and maximises passive solar gains through strategically placed double glazed windows. The hostel is also equipped with a direct drain down solar water preheat system. Two banks of 6.4 m^2 of collectors are coupled via two separate circuits to a 450 litre storage tank. The collectors are mounted on the roof in two banks, one at 45° and one 70° to the horizontal. The system is instrumented and the effect of the different tilt angles on the collector performance is being studied.

Architects: Camphill Architects.

Monitoring: Scott Sutherland School of Architecture, Robert Gordon's Institute of Technology, Aberdeen.

Further information: Saluja, G. S. (1980) and P. Robertson: "Monitoring a solar water heating installation in a 15 person hostel." in **Solar Energy in the 80s.** Conference (M2) Birmingham UK-ISES Midlands Branch 1980, 124-130.

8.29 ELWORTH PRIMARY SCHOOL, SANDBACH

The solar installation forms part of a programme designed to study the performance of different heating systems in primary schools. There are 40 m^2 of aluminium collectors housed in a lean-to greenhouse (which doubles as the boiler house) and are arranged in six banks set at two angles and three orientations, due south, S.E. and S.W. An indirect system using oil as the heat transfer medium is used to heat the 4.5 m^3 solar store. Pre-heating of the kitchen hot water supply is achieved by the immersion of a small preheat tank in the top of the solar store. The conventional system has an oil fired boiler connected to a thermal store. The distribution system used fan cooled units. A system of valves and differential controls sets the temperature flowing through the distribution system at a value which depends on the external air temperature, with water drawn from the water store whenever its temperature exceeds that required in the heating circuit.

Architects: Cheshire County Council, Department of Architecture.

Further information: Tanner, T. W. (1979): ''Solar space and water heating at a Cheshire School.'' in **Practical Applications of Solar Energy in Industry and Commerce.** Conference (M1) Birmingham, UK-ISES Midlands Branch, 1979.

SOLAR HEATING SYSTEM ______________ Sandbach School

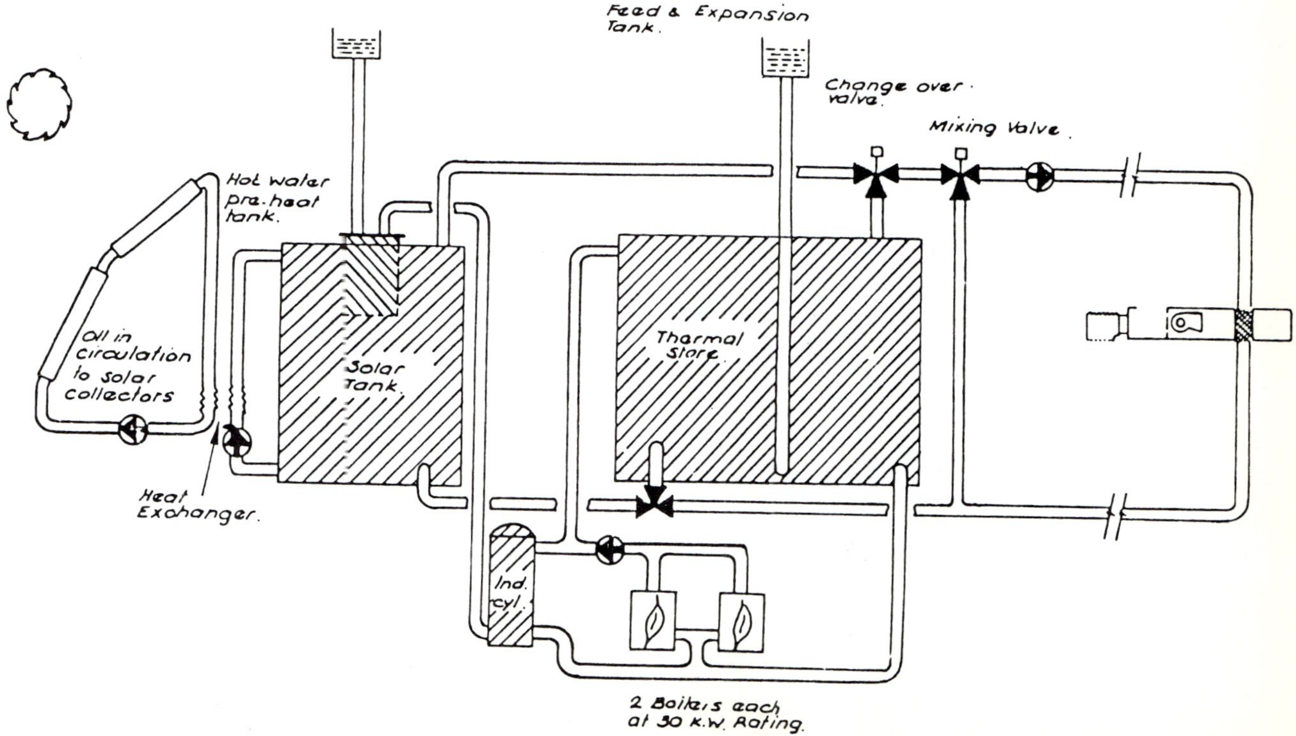

SANDBACH C.P.S.

8.30 **ELDERLY PERSONS HOME, SHAVINGTON**

The two storey 40 person home for the elderly at Shavington, Cheshire, was opened in April 1981. The building exploits passive gains by good design of windows and a massive concrete roof which provides a high thermal mass. The building has 200 m^2 of copper flat plate collector which can feed three storage tanks, using oil as a heat transfer fluid. Domestic hot water is supplied at two temperatures, 65°C for kitchens and laundry, and 46°C for personal use by the occupants. An interesting feature of the design of the heating system is the control strategy. A low pressure aqueous heat distribution system had to be used, with perimeter radiators in the rooms. When the temperature in a selected part of the building falls below a set temperature the boiler is used to heat the building rapidly to an upper set temperature. When this is reached, the boiler is switched off, but the circulation is maintained. When the water in the circuit has cooled below the temperature of the solar store, solar heated water is circulated, until the building has cooled to the lower set temperature, then the cycle is repeated. Thus the solar system serves to slow down the rate of cooling of the building. Solar heated water can be used even at temperatures as low as 26°C, thus maximising the use of solar energy for heating. Computer predictions suggest that 14% of the space heating bill may thus be saved.

Architects: Cheshire County Council, Department of Architecture.

Monitoring: University of Bath.

SOLAR HEATING SYSTEM .... Shavington
Elderly Persons Home

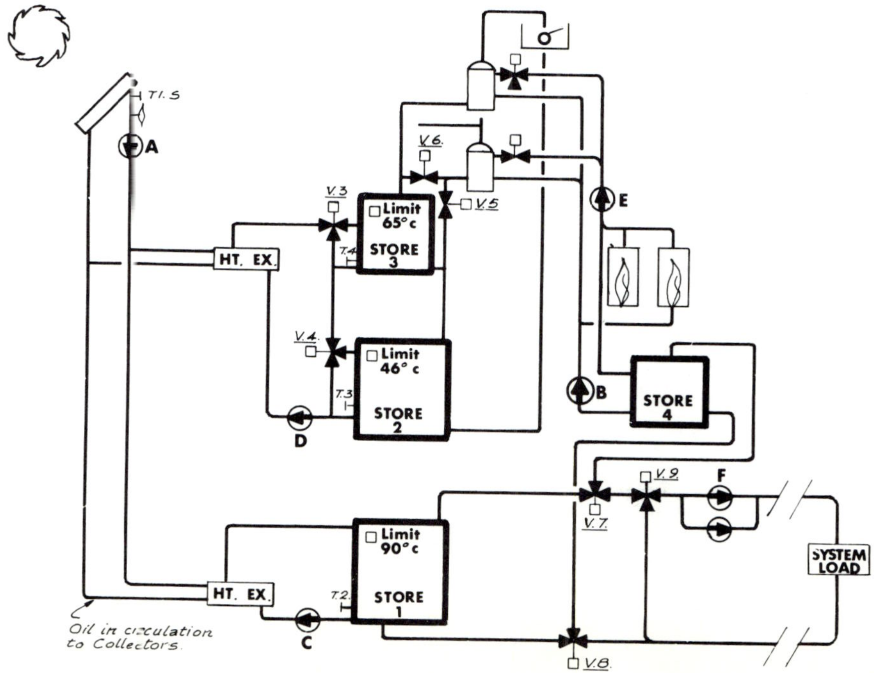

8.31 BRADMORES DAIRY, NEWTON ABBOT, DEVON

Of the £75m spent each year on electricity by Britain's Farming Industry about £35m is used in the parlour and dairy sectors. As part of a programme to investigate ways of reducing this energy bill an energy audit is being carried out, and the possibility of utilising solar energy is under investigation. 10 m^2 of flat plate collector is being used to provide hot water for washing and plant cleaning. Three storage cylinders 190 litres, 280 litres and 360 litres are coupled to the collectors by indirect systems and the effect of different storage volumes on performance is being studied.

Designers: Seale-Hayne College and Plymouth Polytechnic.

Monitoring: Seale-Hayne College and Plymouth Polytechnic.

8.32 SOLAR PILOT TEST FACILITY, BSRIA, BRACKNELL

As part of the CEC programme of solar systems evaluation and computer model validation three pilot test facilities have been established at the Building Services Research and Information Association Laboratories. One of the test facilities, which is common to all the participants in the CEC programme, is a 47 m² flat plate collector system coupled to a 3 m³ water store. This is a wet system designed to represent a typical installation of a single family dwelling to provide space heating. The heat load of the house is simulated by computer programme. The second test facility is a 19 m² mesh type air collector using a 4 m³ heat store. Again the heating load is simulated by computer. A third facility has been constructed to test evacuated collectors.

Designers: Building Services Research and Information Association.

Further information: Potter, I. N. (1980): ''The UK Solar pilot test facility.'' in **Modelling & Monitoring Solar Thermal Systems.** Conference (C24) London, UK-ISES 1980.

8.33 **SOLAR SIMULATOR, UNIVERSITY COLLEGE, CARDIFF**

This simulator, which was the first to use the Thorn Compact Source Iodide Lamp, was designed by members of the Solar Energy Unit, Cardiff, in 1975 with support from the UK Science Research Council (SRC). It has 19 CSI lamps (1 kW) giving a close match to AM2 solar radiation. The solar irradiance on the test area can be varied from 900 W/m² to 400 W/m² and the thermal irradiance from the cooled lamp array is only about 30 W/m² above ambient. The collector and the lamp array move through a full 90° in altitude, permitting studies of tilt and incidence angle effects. Wind speeds can be simulated in the range 0 to 10 m/s. The test area is nominally 2.25 m in diameter, but it has been used for larger collectors such as the Owens evacuated tubular collector shown in position here. The facility is used for research and to provide a collector testing service for industry. More than 1600 hours of collector testing had been completed.

Designers: Solar Energy Unit, University College, Cardiff.

Further information: Gillett, W. B. (1980), R. W. Rawcliffe and A. A. Green: "Collector testing using solar simulators." in **Solar Energy Codes of Practice and Test Procedures.** Conference (C22) London, UK-ISES 1980.

8.34 **ALDBOROUGH, BOROUGHBRIDGE, NORTH YORKSHIRE**

This aerogenerator is the currently available model built by the Sir Henry Lawson-Tancred, Sons & Co. Ltd. organisation. A three bladed, horizontal axis design, it is of fixed pitch construction, which keeps costs and maintenance down. The rotor drives an alternator equipped with a special field energising unit, such that the torque characteristics are matched. Suitable for heating purposes only via a resistive load. The Type 2 machine with its 17.3 m diameter rotor develops 30 kW power at 9 m/s wind, rising to 100 kW at 13 m/s speed. It operates unattended until the wind velocity reaches 16 m/s at which time it automatically shuts itself down. This machine is less expensive than its predecessor: Type 1 which has a hydraulic drive to enable it to generate electricity through a constant speed alternator needed for direct connection to Public Utility network. The Type 2 aerogenerator, shown here, was completed during the summer of 1979.

8.35 BERRIEDALE FARM, ORKNEY

The North of Scotland Hydro-Electric Board installed this machine in December 1980. It is the smallest in a range of four, designed and developed in Denmark. It is called WINDMATIC and has a Horizontal axis 10 m diameter rotor with three braced, fixed pitch blades operating up-wind of the tower and steered by a twin fan tail. Operating experience shows that it can cope with winds in the 5 m/s to 40 m/s range. The rated power of 22 kW is developed at 13 m/s wind velocity. The tower is of sectional design which can be assembled in 6 m units, the machine shown here uses two units giving a hub height of 12 m above ground. During the first winter it has survived gusts of up to 47 m/s (100mph) without damage. The Board is monotoring the operation on site and transmits data to its Laboratories at Pitlochry, Scotland.

Further information: Sommerville, W. M. (1981) and W. G. Stevenson: "An appreciation of the 10 metre Wind Matic aerogenerator operating on Orkney," in **Proceedings of the third BWEA Wind Energy Conference,** Cranfield, April 1981, 120-127.

8.36 BEMBRIDGE DOWN, ISLE OF WIGHT

The most recently installed wind turbine was developed from the original idea of Dr. Peter Musgrove. The "arrowhead rotor" can change the geometry of the blades to suit wind conditions. It is used as a design tool for a 25 m machine being developed by an industrial consortium led by Sir Robert McAlpine & Sons Ltd. The rotor was designed by Aircraft Designs (Bembridge) Ltd. The machine has a 6 m diameter rotor which develops 6 kW in 11 m/s wind. The blades are 0.3 m wide and 4.3 m long. It has operated since 1981.

Designers: Vertical Axis Wind Turbine Consortium.

Further information: Clare, R. (1981) and J. Allan: "Progress with the design of a 25 metre diameter vertical axis wind turbine," in **Proceedings of the third BWEA Wind Energy Conference,** Cranfield, April 1981, 33-39.

8.37 **HIMALAYA MOUNTAINS**

The British Himalaya Expedition used Ferranti photovoltaic solar cell modules to power its communication equipment. They are shown here near the base camp. The protective casing was designed to provide the correct positioning of the array with a tilt angle appropriate to the northern latitude and time of the year. Ferranti are not manufacturing these cells currently.

CREDITS

Numerous individuals and organisations responded to requests for information, facts, opinions and suggestions. Many wrote letters, sent photographs, specifications which are all appreciated, even though some of these have not been incorporated in the book. It would not be practicable to thank everybody separately here so only a brief list follows.

Allen, John
Aspleden, Carl (USA)
Brinkworth, Brian
Burgess, Arthur
Carter, Brenda
Clare, Reginald
Clarke, Freddy
Divone, Louis (USA)
Faiman, David (Israel)
Fritzsche, Albert (Germany)
Green, Cedric
Hall, David
Hardacre, Anthony
Holmes, Barry
Hörster, Horst (Germany)
Hutchins, Mick
Igra, Ozer (Israel)
Johansson, Max
Keable, Julian
Landsberg, Peter
Lebens, Ralph
McGregor, Kerr
Musgrove, Peter
Michaelis, Dominic
Page, John K.
Page, Ken J. R.
Pooley, Derek
Selzer, Horst (Germany)
Spelman, Nigel
Swayne, Roy
Swift-Hook, Don
Tabor, Harry
Telkes, Maria (USA)
Turrent, David

Bournville Village Trust
Building Research Establishment
Building Services Research and Information Association
Calor Group Ltd.
Central Electricity Research Laboratories, CERL
Commission of the European Communities, DG XII (Belgium)
Copper Development Association
Desert Research Institute (Israel)
Dornier GmbH (Germany)
Electra Airconditioning Ltd.
Energy Technology Support Unit, ETSU
ERNO Company (Germany)
Ferranti Ltd.
Ministry of Energy and Electric Utilities (Denmark)
National Council for Research and Development (Israel)
ORMAT Turbines Ltd. (Israel)
Pilkington Brothers Ltd. R & D Laboratories
Robinsons Ltd
Sandia Laboratories (USA)
Sir Henry Lawson-Tancred, Sons & Co. Ltd
Sir Robert McAlpine & Sons Ltd.
Solar Energy Laboratory, University of Birmingham
Solar Energy Unit, University College Cardiff
UK Department of Energy
US Department of Energy (USA)
Wimpey Laboratories

SUBJECT INDEX

NAME INDEX

142